# LIVING WITH WILDLIFE

# LIVING WITH WILDLIFE

David Stephen

*photographs by David Stephen*

CANONGATE

# ACKNOWLEDGEMENTS

The publishers wish to acknowledge the help and collaboration of the Scottish Society for the Prevention of Cruelty to Animals in the making of this book.

They would also like to extend their deep gratitude to Kathleen Anderson, for her many suggestions, and to Julian Stephen for selecting and identifying hundreds of photographs. Particular thanks should also go to Dr James Lockie, David Stephen's old friend and colleague, for his suggestions for the text and help captioning the photographs.

First published in 1989
by Canongate Publishing Limited
17 Jeffery Street, Edinburgh

Text © David Stephen
photographs © Cumbernauld District Council

Design by Gavin Wolfe Murray

British Library Cataloguing in Publication Data
Stephen, David, *1910-1989*
Living with wildlife
1. Great Britain. Animals.
I. Title.
591.941

ISBN 0-86241-215-8

Typeset by C. R. Barber, Fort William
Printed and Bound by Butler & Tanner, Frome, Somerset.

# CONTENTS

Introduction                                           xi

How it all began                                        1

Of Beasts and Men                                      13

Wildlife and Tame                                      27

Rivers, Seashores and Islands                          41

Wildlife in the Past Tense                             57

Clan Weasel                                            61

The Lordly Ones                                        77

I to the Hills                                         93

La Chasse                                             105

Hedgehogs, Moles and Shrews                           117

Reptiles and Amphibians                               125

Hares and Rabbits                                     133

Wildcats and Foxes                                    141

Rodents                                               149

Night Flyers                                          161

To Care or not to Care                                173

A Red Throated Diver at her nest by a lochan in the Outer Hebrides.

*Half-title page.* David Stephen with an orphan Blackbird.

*Frontispiece.* A Rook at nest, well hidden by the early Summer leaves of a Sycamore tree.

*Title page.* A Sparrowhawk floating on an air current.

For my wife – whose compassion for animals
has been her life's pilgrimage.

# PREFACE

When David Stephen was asked if he would write this book, commemorating 150 years of the Scottish SPCA, nobody associated with the project had any indication that it would be his last. In the sad days following his death it became clear that as well as marking the anniversary of the Society, the book had become a testament to his life and to his beliefs. It is clear that he had a strong vision of the future, a future perhaps where mankind finally learned the lessons of centuries of ignorance and came to terms with his environment and that of the animal kingdom. There would have been many times when he despaired, but his optimism, ability to communicate, and his huge personality, have made a significant contribution to our understanding of animals and the world we must all share.

When he died, David had two chapters incomplete, on seals and garden birds. There has been plenty of publicity recently about seals and their plight, not least from the author himself, and we are fortunate to have a fine selection of his photographs of garden birds with their young.

We hope that all David Stephen's many friends and admirers will read this book and remember the man—someone who dedicated his life to animals and to people and who brought a freshness and humour to all those who were lucky enough to have dealings with him.

Scottish S.P.C.A. and Canongate Publishing

*opposite*. A Grey Seal coming ashore to feed her young.

# INTRODUCTION

Greater Spotted Woodpecker at nest in a Birch tree. They are found as far north as Aviemore and on some Hebridean islands in birch and pine woods.

This book commemorates the birth, 150 years ago, of the Scottish Society for the Prevention of Cruelty to Animals. Now amalgamated with the former Glasgow and West of Scotland SPCA, it is the only organisation in Scotland looking after the welfare of domestic farm and wild animals and has been operating for some years Scotland's only cleaning centre for oiled birds at Middlebank Farm near Inverkeithing.

The seal plague of 1988 touched a chord of compassion in many people, among them those who had never given much thought to wildlife or conservation, and the Scottish SPCA won deserved acclaim for vaccinating, then releasing into the sea, a number of seals in the hope that they would survive and pass on their immunity to their offspring. The impact of television was an important factor in stirring popular concern.

The SSPCA, over 40 inspectors and 60 volunteer branches throughout Scotland, is a totally separate body from the English RSPCA. Its aim is to prevent cruelty to animals, rather than prosecute after the event; but when prosecution becomes necessary it needs much greater backing from sheriffs and courts,

*opposite.* Grey Wagtails migrate south in winter and return each April to nest beside rocky streams.

*opposite.* Blue Tit at a natural nesting hole in a tree. Blue Tits and Great Tits take readily to suitable nesting boxes erected where cats can't reach.

*opposite.* The Yellow-hammer is a hedge nester, usually low down. The eggs have characteristic squiggles and lines as markings, prompting a young friend to call it the 'Arabic bird'.

*opposite.* Curlews lay 4 eggs that are enormous for the bird's size.

xi

with higher fines imposed on offenders.

Sadly, the fighting dog is in vogue again, bred specially to entertain spectators with depraved tastes. It took man, the paragon, thousands of years to turn the non-fighting wolf into the fighting bull terrier. The Scottish SPCA regards dog-fighting as barbaric and revolting. It is also unequivocally opposed to badger-baiting and hare-coursing, and works closely with the police throughout the country to catch offenders, who could face a prison sentence if convicted.

One of the many examples of carelessness and lack of thought by owners is those who leave their pets in unventilated cars in hot weather, causing them distress or real suffering. The SSPCA would like the public to think more carefully about animals and their needs, and very carefully indeed before buying one or giving one as a pet. It would like to see much more neutering of domestic pets and Government releasing powers to local authorities for dog registration schemes. It would also like to see ritual slaughter banned.

In its 150th year the society is also deeply commited to the abolition of live animals in favour of a carcass trade, the banning of religious slaughter, stricter codes of practice to improve conditions for the transport of livestock and the phasing out of intensive husbandry.

Contrary to popular belief, not all farmers treat their livestock well and when times are hard the animals are often the first to suffer. The horse is one of the most abused animals with which the SSPCA has to deal.

The work of the Scottish SPCA is vital because, tragically, so many human beings are indifferent to the welfare of animals. To live with animals, responsibly and harmoniously, requires imagination, compassion and, above all, understanding of their natures as fellow creatures. My purpose in this book is to share with all who read it some of the knowledge, insights and pleasure I have gained in a lifetime of study and contact with wildlife and domestic animals.

*David Stephen*
*Palacerigg House*
*1988*

*opposite*. Blue Tit acrobatics. This is one of the few tricks sparrows can't perform.

# HOW IT ALL BEGAN

Feeding a grey squirrel in a Glasgow park.

After 78 years on this mortal coil I am still asked when, and how, I became interested in wildlife, and my answer has always been the same: as far back as I can remember and because I was attracted to anything that moved. Although not born in the country, I uttered my first cries on the edge of it, and grew up with one foot in it. The country was north, the town south, and although I was always one of the boys, doing all the things that boys do, I was a maverick when I headed for the Pole Star.

There were no pets in our house, perhaps because my parents felt that five boys and a girl in the family were enough. I brought home a small puppy once, but had to do an about-turn and take it back to the man who had given it to me. Yet I had three dogs belonging to other people, a bull terrier and two fox terriers, and after school would take one or other of them for a walk. It was after school, too, that I had to do my bird-watching. During the nesting season I caddied at the local golf course and had to do my nest-finding in the evening. My father was a strict disciplinarian who, although a non-church goer, would not allow me to go seeking wildlife on the Sabbath.

It wasn't long after I went to High School that I rebelled over the Sabbath dictum, and my father, while saying that I should find something better to do, relented. I was friendly with our neighbours, the Batemans, who had four boys. Harry Bateman was a motor engineer, who worked a five-and-a-half-day week, but instead of lying in on a Sunday morning was up with the lark and heading into the country. We joined forces, Sunday after Sunday, calling off an early foray only in the face of heavy,

*opposite*. Mallard Drake waddling over a frozen loch in winter.

I

persistent rain. There was a wee bit of head-shaking among the rigidly righteous about the oddness of a forty year old family man going off with a boy at six or seven a.m. and not returning until the middle of the afternoon.

Harry Bateman was good for me; his wife Kate was good to me. He was a jolly man, an extrovert, well-read and good company, and had a fine tenor voice. He was a smoker and liked his pint of ale; my father was a teetotaller and non-smoker. Harry was an Oxfordshire man, yet could recite Burns by the furlong. But for me he had an extra dimension. He loved the country, and was interested in everything in it — wildlife, farming, domestic stock, wild flowers and the changes the seasons bring.

We watched the rooks at nesting time. We found peewit and curlew nests. We watched frogs and toads at the spawning, and reared the tadpoles of both in his house. We watched dragonflies hawking over the pond. We climbed trees to look at the nests of

sparrowhawk, kestrel and owls. We watched foxes at the den. We saw a stoat with a rabbit, trying to drag it into cover and failing. We saw a lot of weasels, and when one dropped a prey we waited on until it came back to retrieve it.

At High School I was fortunate with my teachers, two of whom were of great help to me. My art master lent me a set of bird books, illustrated in full colour, and encouraged me to draw birds, which I did for several years afterwards. My English master became mentor and friend and gave me a number of books which became the beginning of a library. He was a keen amateur archaeologist and told me about the famous Spanish cave of Altamira before either of us saw it. He was Joe Harrison Maxwell, a fine teacher who encouraged his students to write and who sometimes presented prizes from his own library. In the third year I won W. H. Hudson's *Birds in a Village* for an essay on a subject I can no longer remember.

While at school I bought a fat volume called *British Birds* which I paid off at half a crown a month. This, along with the books Joe gave me, were my references. Throughout my school days I spent most of the summer holidays in the country, caddying in the evenings and on Saturdays and going out with Harry on Sundays. I made friends with the local farmers, went to cattle shows with them, groomed and fell in love with the gentle giants — the Clydesdale horses. I worked at haytime and harvest, helped to lift potatoes and shaw the swedes. I learned to milk the cows. I got to know all the poultry breeds of the time. I enjoyed my boyhood.

I remained a maverick except on Sundays. A great day for me was when I found my first nest of weasels in a drystane dyke, and was able to watch the comings and goings of the bitch from a distance of a few yards. She was carrying in voles and mice. I had always been fascinated by weasels and stoats, but that find gave me a sort of fixation, and I have never lost my interest in the whole weasel clan.

I had six years with Harry before I went to Spain with the family to join my father, who was working there. There were six good years of Sundays. My time in Spain, juggling with Spanish, French and English in an office, broadened my horizons. I was in Cantabria, in the mountains, and spent my free days each week following the river or going to the hill. I saw my first wolf there and met the wild pigs, which are easily tamed if caught young. I saw the golden eagle from time to time,

*opposite*. Rooks at their nest in March. The female does all the brooding and the male feeds her at the nest and later feeds both her and the young until they are about 10 days old. He carries food in the pouch under his beak.

buzzards frequently, and red kites every day. I saw foxes often. I knew before I left Spain what I wanted to do, but doubted whether my dream would ever come true. I remain the friend of four generations of one family, who send me cuttings from periodicals that interest me. Although my Spanish is now a bit creaky at the joints I manage to pass muster when I visit them or they visit me.

When I returned from Spain American millionaires were jumping out of windows and splashing their brains on Wall Street sidewalks. Unemployment was increasing, and the world recession beginning. My friend Harry had been killed in a road accident, while I was in Spain, so I was on my own again. Before looking for work I did a month's biking in the Highlands, after which I got an appointment in the Local Poor Law office, which was concerned with the needs of poor people, sick or able-bodied. That suited me because I was then, as I still am, concerned with the causes of poverty and how poor people are looked after.

After seven years I passed my final examinations for my Poor Law Diploma at Edinburgh University and became a member of the Society of Inspectors of Poor of Scotland. Poachers from the surrounding mining villages were among those who passed through my hands, and I had talks with them and watched some of them at work, mostly ferreting or long-netting rabbits. There was also a very old, retired gamekeeper, to whom buses were still charabancs, who left me his notes on pheasant rearing. I got to know doo men and groo men (pigeon fanciers and greyhound owners) and sometimes lectured to the racing-pigeon boys on hawks and falcons.

I had a good deal of free time, and spent all my holidays, annual and local, in the field. The local monthly holiday was Wednesday, so I used to take Monday and Tuesday off, which meant I had four days from Saturday. The Local Authority was very understanding. For some years I was greatly helped by two dear friends, now dead: Willie Dines and James Montgomery. Monty, who ended his war as a sergeant-major in his namesake's 8th Army, was the best tree climber I have known.

The best thing that ever happened to me was my marriage to Jess Russell, who turned out to be a natural at rearing young animals, furred or feathered. She has a great empathy, and is meticulous in her caring. She was still looking after her thirteen-year-old wolves when she had to have a hip replacement at the

Marquis the wolf
in winter at Palacerigg.

age of seventy-seven and I made her slow down. But she still has her interest in wildlife, and I take her around with me every day.

In 1947 I joined the *Daily Record*, which was then edited by Alastair Dunnett, and made a lot of friends among readers, sometimes giving short, informal talks to men at their workplace during the lunch-break. Shortly after joining the *Daily Record* I quit my job with the Local Authority and bought a farmhouse, complete with byre, stable, hayshed and a bit of ground, and was at last free to do what I wanted to do. In 1956 Alastair Dunnett became editor of the *Scotsman*, and asked me to join him.

My wife, son and daughter loved the farmhouse, which was beside the Luggie Water a stone's throw from the village of Luggiebank. Soon after we were settled I bought a Border terrier bitch puppy from The MacLean at Kingshouse across the road from Buchaille Etive. My big Labrador, Fencer, took charge of her right away, as he did later when I presented him with a seal-point Siamese kitten. I reared a flock of Light Sussex pullets, added bantams, geese and ducks, and the following year reared pheasants and partridges. Foxes I had to have, and I reared three. Then I began breeding Large White pigs. And I had a tame raven, born Atholl.

Part of the byre I used as a sick bay which, over the years, housed a golden eagle taken from a gin trap, a wounded peregrine falcon, owls including a short-eared that flew to my wife's fist, roe deer, hedgehogs, badgers, a shot-up hen-harrier that died, roe deer fawns, a gannet and a whooper swan. Indoors my wife cared for a motley throng, including a young cock pheasant called Hamish and an American sparrow hawk that had been picked up on a weather ship 400 miles out in the Atlantic. When it was fully fit again it was returned to the United States on a Flying Fortress out of Prestwick, authorised by a senior officer of the United States Army Air Force, who made so little fuss you'd have thought *Falco sparverius* was a regular passenger.

Raven with Golden Labrador; Ravens love to tease cats, dogs, in fact any domestic animal if they can get away with it.

We had nine wonderful years at Luggiebank. I had the freedom of all the surrounding farms, and three estates, only one of which was keepered. Nothing would have persuaded my wife or me to leave the farm, but leave it we did in 1958 — not by persuasion but by order: compulsory purchase. So out we had to go. So had the bats. And when my ringed swallows returned the following spring there was no inn left for them to find room in. We left some dear friends behind — in the ground: Susan, our pet sow; Fencer the Labrador; Bounce, the roebuck; and Tosh, the half-bred wildcat.

We were fortunate in being able to find a new place before we had to quit the farm. Three bachelor brothers, whose villa and ground I had admired since boyhood, learned of my predicament and sent me word that they were leaving to live in a house that had been built for them in Haddington. Would I like to buy Firknowe? I would. And I did. It was all agreed within an hour, and sewn up by the lawyers within a few days. Firknowe was only two miles from my old place, so I was still on my stamping ground. But I had to write off my ringed swallows and the bats. Firknowe had neither.

My old friend Jim Lockie, whom I've known for more than thirty years, came to visit me when the house was being torn apart and renovated. I met Dr Lockie when he was a scientific officer with the old Nature Conservancy, working on stoats and weasels in the Carron Valley where there was a vole plague. In 1955 I worked with him on an eagle survey in Lewis, and we have been friends ever since. I told him of my plans. I was going to have stoats and weasels, polecats and foxes, badgers and wild rabbits, all at home, so that I could study them captive as well as wild. My hope was that if I had them tame I could follow them afield as well. It was a tall order, but some of it was to come to pass. Lockie was to leave the Nature Conservancy, to become a lecturer at Edinburgh University, and thence to Peebles to make hand-looms. How's that for a scientist of many parts?

I forget the precise order in which I did things over the next two years, but I finished up with a walled garden (I built the wall myself, with a door into the spinney where I had my poultry), an extra outhouse in the north-east corner (the wall made two sides of it), round-the-walls roomy cages for stoats and weasels, a pool for water voles, and a big enclosure for foxes. I live-trapped a pair of weasels, got a pair of stoats from Lockie, reared fox cubs, netted wild rabbits for my outhouse, and later on installed a pair of badgers.

Skipper hunting voles on a
hot day in summer.

My dogs were the late Fencer's daughter Tarf, and the Border
terrier Nip. We had by then three cats: Satan the seal-point
Siamese, Snowy the white neutered female, and a powerful
neutered male called Cream Puff, because he was all cream. But
he was no puff. He could hold and kill a six pound hare. Tarf,
like her father, became blind at the age of seven, but was able to
operate her home ground by kinaesthetic sense.

We had more than the usual number of transients, the injured
and the 'rescued', perhaps because Firknowe was more accessible
than the old place. A roe-buck, injured on the road near Kilsyth,
was brought to me by the police. He had head and body
wounds but no broken bones. Within twenty-four hours I

realised he was blind, so I called in my vet, who said there was a good chance that his sight would come back. He was right. A fortnight later he had full vision and I turned him loose in a wooded glen belonging to a friend of mine.

Later in the year I released an old roe doe in that same glen. She had been caught up in barbed wire near Rutherglen, and some boys had peppered her with airgun pellets before they were chased away by an adult who got in touch with the police. I stitched up her wounds and got some pellets out of her hide, and released her when she was fit. For two years I saw her from time to time, recognising her easily because she was such a ragamuffin and had a slight limp.

Early one April morning a man arrived with a sow badger that he had found in a fox snare half a mile from my house. The wire was deeply embedded in her hide and it took me some time to get under it and cut it. She was a milker, so she had cubs, which I reckoned would be all right. I released her where the man had found her, and watched her shamble off to her sett less than a hundred yards away. I knew the sett well and when I watched it that evening three cubs came out, followed by two adults. On the way home I caught my right foot in another fox snare and took a dive.

Caring for the sick, the injured or the maimed, was something my wife and I undertook as a matter of course, and when killing was necessary I did it myself. What to do with young wildlife that should never have been brought to us in the first place was a perennial problem. When a roe fawn or a small leveret, or an owlet, was brought to us within a few hours of having been picked up we usually managed to put it back where it belonged. But if it had been in somebody's care for two or three days, or passed from one person to another, we would rear and hope for the best. Tragedy is a common end for such wildlings reared tame.

During my years at Firknowe I lectured on wildlife for Glasgow University's Extra-Mural Department. I had two classes each winter, and during the summer I arranged excursions with them to various parts of the country. One day was set aside each summer for a visit to Firknowe, where we spent the time looking at my animals and going round my box traps, of which I had twenty-four on 800 acres. When I produced my tame weasels, Tamma and Teen, which I had reared on the bottle, they became the star attraction, running upstairs and down, up

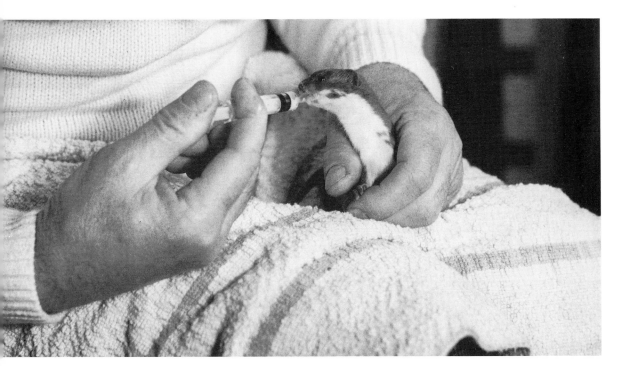

Big Dave feeds wee Tammas the baby weazel with milk from a hypodermic.

curtains and down, up legs and down, up Tarf and down. The pair had an upstairs room in the house, and adjoining roomy enclosures in the stable. People from the village, and sometimes passers-by, used to stop on the road to watch them playing up and down the curtains in their room. Carolyn King, who was researching weasels at Oxford came to visit them and spent the afternoon with them in their room. She found them enchanting and wanted to take them to Oxford, where they would be treated as 'honoured guests'. My wife and daughter said No. I said No. We had cause to regret that decision bitterly the following spring, when we all moved to Palacerigg Farm.

Cumbernauld Town Council, as it then was, invited me to make a Country Park out of this bleak moorland farm, and after some thought, and several friendly talks with councillors and the Town Clerk, I accepted. I was to become Director of the project on a part-time basis, remain self-employed, and work directly with the Council. I knew the ground well, having worked all round it, and sometimes on it, since before the war. I reckoned it would take ten years to get the place established. In the event it took a little over eleven years.

It was agreed that I should continue to farm enough ground to make the Park self-supporting in hay and oats for domestic stock, all native breeds: Highland and Shetland ponies, five breeds of cattle (Highland, Ayrshire, Aberdeen Angus, Black and Belted Galloway), a small unit of Soay sheep and one of 'wild' goats. Besides these we would have red, roe and fallow deer, ducks and poultry. It was also agreed to put on show foxes, badgers, stoats, weasels, polecats, wildcats and mink. I would bring all of these from Firknowe except the wildcats, but my wife would rear a pair of kittens the next year. At the same time I would rear grey lag and Canada geese, and mallards, all of which would be left free flying and breeding. No birds, other than sick or injured, would be grounded or caged.

When I moved to Palacerigg in February 1971 my house was ready but the animal enclosures were not, so all of them had to be given temporary accommodation in a small byre. I saw them several times a day so they didn't fret much. Vandals broke in and beat them all up one Sunday when my wife was giving lunch to a team of voluntary workers in the house. They killed my stoats, the weasels Tammas and Teen, all but one of my polecats, and the mink. Spick and Span, my badgers, were safe. So were the foxes Glen and Fiddich. I found Tammas later in the day, dragging himself back to his bed with a broken pelvis. I had to kill him. A third fox, which my wife had reared the previous year, escaped only to be shot before he had travelled a quarter of a mile. The killers were found, charged, and fined heavily. Jess and I often live that Sunday over again.

Kathleen feeding one of the foals at Palacerigg.

# OF BEASTS AND MEN

Mallard ducklings reared by Kathleen.

People often ask Jess if she has, like Dickens, in her heart of hearts a favourite among the many wildlings she has reared over 50 years. Before 1975 she would have had to think before answering, if she had an answer at all. She doesn't have to think now, because in that year her favourites were born — her big Alsakan timber wolf Marquis, who is still her pride and joy, and his mate Magda.

One day, when the wolves were about five years old, a woman with a visiting WRI group asked her: 'Mrs Stephen, are you not afraid that big wolf will go for you one of these days?' When she said she wasn't, the woman then asked: 'But how can you be so sure?' To which Jess replied: 'Because he's a wolf!' In four words she encapsulated all modern wisdom concerning the wolf. My daughter has been feeding them since February 1987 when Jess had a hip replacement.

Her second favourite, perhaps even joint first, was her greylag goose Polar. Although the egg in which he developed was incubated by a wild grey-lag in south-west Scotland, Polar

*opposite*. Jess with Marquis the wolf at Palacerigg Country Park in winter.

13

hatched out on a sheet of felt spread on the kitchen table. From the beginning Jess saddled the bird with the masculine gender, and when she called his name he came to her. She took him out to graze, and to swim and dive in a pool. I was out every day, for most of the day, and he spent some part of each walking with me before he could fly, and even after he could fly. But sometimes he flew above me, circling and touching down, then taking off again. I could call him down from the sky any time I wanted to, and usually did when schoolchildren asked me to 'Ca' Polar doon.' Then they would fuss him while he spoke goose talk to them.

He knew my van, and one day when I stepped out of it in the town centre I found him on top of it. I put him inside until we were clear of town, then let him out, and I had a flying escort home. At the official opening in 1974 of Palacerigg Country Park he stole the show. The late Sir Charles Connell, then chairman of the Scottish Wildlife trust, was the official opener. While the press boys were posing Charles and me for a photograph Polar crash-landed on my shoulder. The *Scotsman* photographer caught the moment and the paper carried the picture next morning.

Polar the Greylag Goose, Lisa the German Shepherd and David Stephen.

In March he became very excited, running in and out of the kitchen 20 times a day, gabbling incessantly to Jess and me, and I told her that he was ready to go and wanted us to go with him. I suggested grounding him until the moult, after which he would fly again and almost certainly stay with us for life. But she refused to see her beautiful goose made flightless, even for a few months. She was also certain that he wouldn't go. But go he did. She consoled herself with the thought that he might find and take up with his own kind, an unlikely event because he ignored his own kind at home. I, from my seat in the real world, thought he was more likely to come down to earth to pass the time of day with some person, another Jess perhaps, but just as likely to be some cowboy with a gun.

While I was operative at Palacerigg I had a phone call from a woman telling me that there was a kestrel hanging by its jess strings from the electric cable outside her house. I drove to the place right away and saw a female kestrel on the wire, now right way up, now upside down, with her jesses tightly fankled. The first engineers to arrive thought they could pull the bird in using a pole with a hook, but I disagreed, saying they were more likely to break her legs. Switching off the transformer and bringing the cable down was out of the question, they said. But a more senior person from the south of Scotland Electricity Board arrived who, at my request, did just that. When the cable was lowered the bird ended up in a tree, which I had to climb to cut it free. I kept it for a couple of days, during which I fed it all the mice it could eat, then set it free. I gave full marks to the SSEB for their prompt and full co-operation.

I have long enjoyed the co-operation of the police for miles around, and, always enjoyed co-operating with them. A good example of bold decision was when a rocky bank had to be blown. It had been blown before, to prevent children from crawling into a section of old fireclay workings, but they were still able to get in, so it had to be blown again. But there was, for me, a problem. Badgers were denned under the rocks, and I had seen five coming out on several nights during the week before the explosive charges were to be laid. I told the explosives men that it was not the intention to blow up the badgers, and they said there was no alternative. There was. I asked the police if the site could be blown at night, after all the badgers were out, and they agreed. But how would the explosives men know when to blow? That was easy. I would watch until all the badgers were out, then give them a signal. And I did.

Then there was the badger in the man-hole. The police told me that the cover had been removed and the badger had fallen in. The man-hole was deep, with several inches of cold sludge at the bottom, and the badger looked like a wet plaster sculpture. I brought a tall set of kitchen steps and lowered them into the man-hole, then put a noose round the badger's neck and walked it up. When I was removing the noose the beast made no attempt to bite me. It staggered a bit when it shambled away to the sett in the glen. It would almost certainly have died of hypothermia that night if the police hadn't noticed the uncovered man-hole.

Farming has changed greatly since the War and so have many farmers, but there are still many of the modern breed concerned with more than the pursuit of the fast buck up grain mountains or across milk lakes. One such was a friend of mine who had forty-two pairs of house martins nesting under the eaves of one of the farm buildings. After one nesting season he had the building re-roofed, which meant that nearly all the nests were destroyed, and he was concerned about how the birds would react when they returned. The birds started to build new nests, most of which kept falling down. Was this because of the new wood or the wrong nesting mix? The farmer delved over a patch inside the field gate, spread well-rotted dung over it, added a sprinkling of cement, and stirred the lot into a thick porridge. The birds used the mix for their nests, and the nests stayed stuck.

One of my neighbours when I was at Firknowe was the late John Carmichael, vc, who had barn owls nesting at his farm. He gave me every facility when I was watching or photographing them, even to the extent of running a power line into the gable doocot where they lived and nested. The time came when great alterations had to be made to the building, which would mean the owls losing their doocot, and John was anxious to know how he could keep them. I suggested fixing a very big wooden box to the inside of the opposite gable and giving the birds access to it via a hole in the masonry. He liked the idea, had the work done before the owls lost their home, and they moved into the new accommodation.

Another farmer friend, having bought a braw new car, decided to keep the usual pair of swallows out of the garage from then on. There were many swallows at the farm, and no

Barn Owl in nest with her owlets in the attic of a barn. The nest is just a hollow in the carpet of owl pellets.

scarcity of other nesting places. So the garage door was kept shut. And then a strange thing came to pass. One morning the farmer discovered crumbs of nesting material on the roof of his car, and saw on an overhead rafter the beginning of a nest. The birds were gaining access to the garage via a small hole at ground-level where a bit of the door had rotted. The farmer nailed hessian to the underside of the joists to catch droppings and debris and opened the door.

In 1968 I was introduced to a retired farmer from Essex, an expatriate Scot who had left his native heath as a boy half a century before. We got to talking about partridges, and about how scarce they were becoming, and he told me a story, the telling of which obviously gave him great pleasure.

'Afore I retired there was this day I was looking at the hay, bendin doon ye ken, when suddenly there was this twa shotes, and twa pairtritches was doon — wan o the cowboys fae a neeborin ferm. So I lookit aboot an picked up maybe ten or sae wee cheepers, an took them hame in ma bunnet, an ma dochter Margaret took them, and lookit efter them, an they cam on good, an flew, but widna gang away. They stayed on an next year some o them were still there an they bred an syne were gaein aroon wi their cheepers. An it was a grand sicht, an I just thocht the place widna be the same if a thur things were gane.'

And I said to him: 'Man! When I was in Dublin's fair city a while back, speaking to schoolchildren, a Dubliner said to me: "Wouldn't the world be a hell of a place if it was filled with nothing but people."'

A farmer whom I had known since boyhood never allowed roe deer to be killed on his ground. When I came to be a neighbour of his he phoned me one day to say that a roebuck had hustled one of his in-calf, pedigree heifers into a pool where she had almost drowned. Why should a roebuck do such a thing? All I could suggest was that he must have been a beast ousted from his territory, with no doe and nowhere to go. I based this opinion on my own experience of having a tame yearling buck mount my knees when he wasn't trying to mount the dogs. He said he would have to shoot the buck if it bothered his heifers again. I told him I thought this was a one-off event, not likely to occur again in his lifetime or mine. The buck didn't repeat the exercise and the farm remained a no-go area for people wanting to stalk roe.

That farmer — he has been dead these many years — had a feeling for wildlife, and a saying that a 'wee bit of trouble, at nae cost' was often a help to something, like lifting and replacing the eggs of peewit, curlew or oyster-catcher during

Roe deer in velvet. Deer take good care of their antlers at this stage as they are sensitive and easily damaged.

field operations, or running the dogs through the last stand of hay before cutting in case there was a roe fawn, or fawns, lying up there. One morning, at first light, he came out to find a stoat lapping milk from the cats' basin. I was in the hayshed, ten feet or so from the ground, watching too. When I spoke to him he called to me: 'Five unembliddyployed cats primpin and coyfurin while a lousy stoat-wheasel steals their milk.' The cats were, indeed, licking their fur, but they had left the milk basin some time before the stoat arrived, and I told him so. But he was angry with the cats for what he thought was their surrender, which was a nice bit of ambivalence because stoats and weasels on his farm were never molested, and every time he saw a weasel he would call it a *Johnny Whittret.* Once, at a threshing, he bawled at a worker who had taken a swipe at a weasel with his pitchfork: 'Leave it be! Whit the hell dae ye think it's daein? Eatin corn?'

One day I called on a farmer to ask permission to photograph a curlew nesting in a big field of rough grazing. When he told me that he would be putting twenty or more stirks on it at the weekend I called it off because cattle had wrecked hides of mine before and I wasn't inviting a repeat. How long would I need? Four or five days, I said. The man agreed not to put the beasts

Curlew on nest; Curlews nest on hills and moors and in winter move to the seashore. The trill of the first Curlews, back in the hills is the sign that Spring is nearly upon us.

in until Monday. I asked him if he would drop off eight or so fence posts and some barbed wire, and I would fence the bird off in case the bullocks trampled the eggs when they did their cavalry charge on arrival. The man said that would be nae bother, and the tractor man would help me put it up.

I visited the curlew late on the Sunday and decided to spend the night in my hide, a thing I've often done. The bird was asleep on the nest when this heifer broke through the fence and came thudding up to my hide like Don Quixote charging a windmill. She dunted my back, and I elbowed her in the face. Then I looked out at the curlew and saw that she was now standing, half crouched, over her eggs. The heifer dunted me again, and I told her in a low, fierce voice to get the hell out of it. The curlew raised her head, and lent me an ear, but settled again almost at once. The heifer dunted me again and I elbowed her back.

This time she thudded away a few yards, and stopped. The curlew lifted from her eggs, a shadowy shape in the ground gloom, and pitched a few yards on the far side of the nest. If the

David Stephen's 'hide' beside a lochan in the Hebrides for photographing Red Throated Divers.

heifer left now I was sure the bird would come back immediately. Instead, she came back while the heifer was still swithering. She straddled her eggs, lowered herself slowly, shuffled her wings, and settled, pressing beak to shoulder. The heifer butted the barbed wire, didn't like it, and drew back snorting. Then she walked round the fence, prodding it lightly, seemingly more interested in the sitting bird than in my hide. But, presently, she came at it again, determined to wear it on her head and we had a boxing match of five-seconds rounds with five-minute intervals. And all the time the curlew sat on, lifting her head now and again in query, but not taking fright or flight. It was an hour before the nuisance shambled off, heading for the hole in the fence she had broken through earlier. A week later the curlew hatched four chicks.

Some birds will put up with a great deal of unavoidable disturbance if the disturbers show some consideration for them and minimise it. Mobile nests are a good example. A farmer, whose tractor has been idle for a fortnight, suddenly needs it and finds a pied wagtail sitting on a nest on the cylinder block. She flies off when he lifts the bonnet, but is back within a minute of his closing it. And where do we go from here? the man asks himself. He thought she might desert when he started up the machine, but the engine didn't bother her. He needed the tractor, but he streamlined his work programme, adjusting this

Pied Wagtail feeding young on the ledge of a barn.

and that and bringing in his other tractor. Even so he needed the one with the nest for an hour a day, half an hour out and back morning and evening. The bird took the half-hour off while the tractor was away and went back to work when it returned.

Once the wagtail had hatched out the tractor was left idle except for brief, unavoidable use. The birds accepted the interruption to their feed-routine and reared their brood. The farmer lost a lot of tractor hours, but got a lot of pleasure, which can't be put in the bank; but, as he said, 'There's mair tae life than the fast buck or the slow wan.' And that's about all of Philosophy when you come to think of it.

A mistle thrush built her nest one year on the jib of a huge crane in a scrapyard. The nest was built, and the eggs laid, while the crane was idle, but the bird had to incubate while it was operating, swinging this way and that, lifting and dropping the great ball of metal used for breaking up scrap. After the young hatched the parents were faced with a feeding problem. It is one thing sitting in a mobile nest, but something else to fly after a moving jib with a beakful of food. The birds solved the problem by coming to one spot and sitting there until the jib swung round to them. Then the operator stopped for a few seconds to let the parents feed their young and clean the nest. It was a great success story. Nobody lost any time and the men were all proud of their mistle thrushes.

I knew a pair of blackbirds that had their nest on a digger in a sand quarry. Unlike the mistle thrushes they didn't wait for the nest to come to them at feeding time; they flew to wherever it happened to be and fed their young on the move. This was a big adjustment for the birds to make, because a nest is a very restricted spot on a map and they have a very restricted mental map of it.

A priest came to me once to help him with a problem. A blackbird was sitting on eggs on the engine of his curate's small car, and the curate was due back from holiday the next day. Had I any suggestions? That was easy. The curate could remove the nest, forcing the blackbird to build another somewhere else, and he would have instant use of his car. Or he could let her hatch out and rear a brood, in which case he would have to do without his car for two or three weeks. The priest looked me straight in the eyes — he was old and Irish, and the look was old-fashioned — and made his decision: 'Och well,' he said, 'the walk will do him good!'

Gamekeepers, generally, are not my favourite people, but there always have been top-liners among them, the cream on top of the milk. Top-liners are not thick on the ground, but easy to identify when you find them. One such, whom I have known for nearly forty years, had sparrowhawks nesting with him at the time when the species was in decline. The nest was close to the boundary line, on the other side of which the keeper killed anything with a hooked beak. When the brood of six young sparrowhawks was in the branches my friend and I climbed trees, trying to shoosh them away from the danger area, and we maybe managed to double their distance from the boundary. But it was an exercise in frustration, because there was no way of keeping them there, and more than likely they would be at the nest again soon after our backs were turned.

Can you imagine a keeper, who was also the owner of a small flock of breeding ewes, keeping a golden eaglet alive after its mother had been killed? You don't have to imagine it. The man was a friend of mine and I was working at the eyrie concerned. He and I fed the eaglet night and morning on the prey brought in by the cock, and it quickly learned to tear up food by itself. That eyrie was special in another way. One morning we arrived to find a lamb in the eyrie, bleating to its *baa-ing* mother on top of the crag. It had squeezed into my hide, then out in front of it, then leaped the narrow space between it and the eyrie. And there it was jumping about, and sometimes trampling on the eaglet, making it *peep-peep* in protest. My friend got his crook round the lamb's neck and deftly hooked it on to the narrow ledge in front of my hide.

Then he addressed the eaglet thus: 'By cricks, wee *Chrysaetos*, just fancy your breakfast coming and gie-in itself up!' Then to me he said: 'By cricks, Dovit, it's a good thing I was here wi ye the day. Naebody wid believe it! I can just see the headlines in the paper: Lamb kills *Golden Eagle*!' Keepers like these are a pleasure to work with, and I've met them here and there from the central Highlands to the Outer Hebrides.

We humans are an untidy lot, especially the litterbugs who mess up the countryside, yet Nature sometimes manages to make use of some of it, like great tits rearing a family in a two-gallon petrol can, or badgers tearing the stuffing out of a sofa and dragging it away as bedding, or a roe deer walking about with a syrup can on its muzzle, or a fox sleeping in an armchair, or voles nesting in old bedding. I once saw a cow with its head in a picture frame. But there is the serious, as well as the unsightly, side. My big German Shepherd bitch almost lost a foot when she trod on a broken bottle in a pool. A meal of plastic doesn't do cattle or deer much good. Nor does it do fish any good when somebody throws a lidless, supposedly empty Cymag can into a pool. I used to pick up some dumpings in my van, and take them to one of the local tips, but not any more. There are so many of them, some big enough to bury the *No Tipping* signs, that I would have a full-time job.

The genius of Leonardo da Vinci included the gift of prophey, and on the subject of beasts and men he wrote:

> The human creatures will always be fighting each other, with frequent death on either side. They will destroy the vast forests of the world and when they are filled with food they will deal out death, labour, terror and banishment to every living thing. Nothing on earth or under it, or in the water, will not be pursued, disturbed, or spoiled, or removed from one country or another.

Our own Charles Murray had the answer in his poem 'Gin I was God':

> To some clood-edge I'd daunder furth, an feth,
> Look ower an watch hoo things were gyaun aneth.
> Syne, gin I saw hoo men I'd made mysel
> Had startit in to pooshen, sheet an fell,
> To reive an rape, and fairly mak a hell
> O my braw birlin earth — a hale week's wark—
> I'd cast my coat again, rowe up my sark
> An, or they'd time to lench a second ark,
> Tak back my word and sen anither spate,
> Droon oot the hale hypothec, dicht the sklate,
> Own my mistak, an, aince I'd cleared the bord,
> Start a'thing ower again, gin I was God.

*opposite.* Golden Eagle chick about a month old and about the size of a White Leghorn hen.

A bit of overkill, perhaps, but a little more humility and caring on the part of man, the conquering hero, would help.

# WILDLIFE AND TAME

Tara, the Jack Russell, sharing her basket with a young hare.

Relationships between wildlife and tame are often obvious enough, like cat hunting mouse, dog chasing rabbit, fox in the henhouse, crows killing poultry chicks or ducklings, or where ground-nesting birds and hares sit unalarmed in a field where big domestic livestock are grazing, yet flee at the sight of a human being, or a pack of beagles, or any dogs, or one dog. The hare sees man and dog as equal predators.

Man is responsible for creating many of the contacts between wild and tame, like foxhounds hunting foxes and staghounds hunting deer. Otter hounds used to hunt otters until otters became protected by law; now they hunt mink instead. Beagles and harriers hunt hares. So do lurchers and greyhounds. Terriers are used for bolting foxes, or in the obscene 'sport' of badger baiting. Ferrets are used to bolt rabbits into purse nets, and dogs to drive them into long nets. During illegal badger digs terriers are used to keep the brock on the defensive so that it can't burrow its way to safety. Terriers are also used at rat hunts, and were a familiar sight in farm stackyards at threshing

*opposite.* David Stephen training Kirsty the black Labrador. An earlier Labrador, Tarf, was blinded by eating strychnine laid in a bait but he lived on to the age of 16.

time in the days before combine harvesting. From all this it can be seen that man uses the domestic mostly to hunt and kill the wild. He will even use tame to kill the tame, as in the case of cockfighting or dog fighting, both long outlawed but increasingly in vogue again.

Yet many people often do the opposite, creating situations where wild and tame, even when they are natural enemies under natural conditions, can be brought up to live amicably together. I have had cats that slept with owls, and had their ears pecked by them without protest, I had a tame woodpigeon that played with the cat, which often grappled with the bird without unsheathing a claw. Fencer, my Labrador retriever, and Lisa, my big German Shepherd bitch, adopted anything brought into the house, from roe fawns and leverets down to ducklings and baby weasels, and photographs of such associations are ten a penny. I once said to Clive Hollands of the Anti-Vivisection Society that such photographs in Annual Reports of this or that organisation tended to make people think that such would be the situation if only man would keep his mug out of the picture. The opposite, in fact, is the truth. Without man's contriving cats wouldn't suckle baby rabbits, any more than my terriers

Liza with the wolves, Marquis and Magda.

would put up with leverets in their baskets or weasels nibbling their nails.

I have always been fortunate with my dogs, from Labrador Kirsty to dear blind Tarf, who died at the age of 16; from German Shepherd Lisa, who mothered wolf pups, to my present German Shepherd Shona, who would rather research mice than hunt them. Lisa was very special, another Jess in dog skin. All my terriers, Borders and Jack Russells, were allowed to kill rats and rabbits, but nothing else. As for the cats: we reared all our own. They were taught the facts of life from kittenhood and knew, as grown-ups, that rats, mice and rabbits were fair game, but that not hunting birds was Canon Law.

Cat persons who say that you can't teach a cat to leave birds alone are really exempting themselves from trying although, generally speaking, cats and birds don't mix, so anyone who feeds wild birds through the winter should make sure that feeding stations are out of their reach. My cats would sometimes spit at Spider, the tawny owl, if he nipped them too hard, but that was the limit of their riposte. Our Siamese cat, Skipper, used to lie under the brooder lamp, with tiny pheasant chicks running over him like fleas. At our farmhouse I used to leave Cream Puff overnight with small poultry chicks until we had finished off the rats in the out-buildings.

Any time we had a small leveret in the house the cats took as much interest in it as the dogs did, seeming to understand that, although it was furred, it was not fair game indoors. They were the same with a budgerigar I caught in the garden. We couldn't find its owner, so we held on to it, although Jess and I are not great budgie people. We gave it a cage and let it fly about the house, much to the annoyance of the cats, which kept seeking refuge from its overtures. Jess reared a lot of small birds in the kitchen, including a whole family of blackbirds, without ever having to protect them from the cats. Indeed, more often than not, the cats got out of the way of owlet, kestrel, magpie, jay or whatever.

Feral cats are a different matter and you won't teach Canon Law to one of them if you give it a home. In 1987 I rescued my resident cock blackbird from a feral cat. The cat got his tail and injured his left wing whose elbow remained thereafter at the half-salute. But the bird survived, and kept coming to the kitchen window-ledge for raisins. When his tail grew in again it had two silver feathers. He lost his new tail too, presumably to another, or the same, cat and disappeared not long afterwards. One of his sons, almost as tame as his father, now comes to the kitchen window for raisins. When I arrive first I call him; if he sees me first he calls me.

They'll tell you that a pheasant is wild from the day it flies, which is generally correct, but not if you take a day-old chick and rear it by hand. I found one such mite dying in a ditch, but instead of executing it with my thumb nail I carried it home inside my shirt. The bird grew into a handsome ringneck cock and we called him Ringneck. My Labrador retriever Kirsty looked after him when he was wee, and running about the garden, and they remained close friends for the rest of his life. He flew round and round the houses, walked with the dog, fed in the fields, and once created a traffic jam three-quarters of a mile from home. The police, who knew him, managed to shoosh him into the air and told me about it later. One day, in October, he paid a visit to a fishmonger's garden, a few hundred yards from home, and was shot there. The fishmonger, seeing the ring on each leg, realised whose bird he was and was sorrowfully apologetic. Yet you'll still hear that sportsmen don't shoot sitting birds.

I've already mentioned our billy-goat William, who chased a fox from his field. One August he had a head-to-head argument with a roebuck at the fence. But the most incredible contact between a roebuck and a billy-goat that I have ever seen surely has to be classed as a one-off. I was on a visit to the big pleasant Irishman who kept a motley crew of goats in a field behind his house. I don't know how it all began, but when I came on the scene the buck had one antler under the goat's collar. When I ran to them, intending to cut the goat's collar, the buck broke free, at the cost of a broken antler. Years later I used the incident in my book *Six Pointer Buck*.

During my Luggiebank and Firknowe days I spent a lot of time in the farm fields at night, and was greatly interested in the behaviour of cows, ponies or sheep towards visitors from the wild. Cows, more often than not, would follow a fox in mass formation and drive it from the field, but sometimes a bullock would leave its fellows and give the reynard a real rousting. The usual reaction to a passing badger was to stand and stare. But they were familiar with the badgers, whose setts were nearby and whose trails crossed parts of the fields. A roving dog would get them to their feet, and one herd, in the field flanking my garden at Firknowe, always harried my terrier to the fence. At Luggiebank my neighbour's beasts sometimes chased Bounce, my tame roebuck; at other times they did no more than turn their heads slowly to watch him slouching by. The farm cats they ignored completely, presumably because they recognised them as fellow dwellers in the byres. They weren't so tolerant of our big Cream Puff if he took a notion to go mousing on their side of the drystane dyke.

At Luggiebank we had two white Saanen nanny-goats, whose pastime was running head-down at any fox they saw. My

William, the Wild Billy goat.

neighbour's sheep greeted any fox with a stamp of a foot, and I've seen a ewe, with a lamb at foot, put a prowling fox to flight. Ponies always seemed more tolerant of wild visitors, although Highland foals would follow a fox out of curiosity. The same foals would run to their dam at the sight of a strange dog. A hare lolloping into their field attracted their curiosity, and I remember one that allowed a foal to approach almost to sniffing distance before taking a few unhurried hops out of range of the questing muzzle.

Our Shetland stallion Naughty Boy — gifted as a weaner to Jess by her dear friend the late Viscountess Templetown — was a horse of a different stamp, and a great wildlife researcher. Until he was a year old his favourite plaything was a lightweight wooden plank, which he would carry about, balanced in his mouth, or toss into the air to be clobbered with his hindfeet on the drop. One day a carrion crow took an interest in his capers, which obviously displeased him, because he dropped his plank and charged it until it swept up to a telephone pole from which it sent crow-talk to his ears. Naughty Boy returned to his plank and did a kind of war dance on it. From that day he was a crow chaser, and rook, carrion or jackdaw were all the same to him.

Eriskay ponies on the island of Eriskay in the Western Isles. They were soon to be moved to Palacerigg.

Yet he would tolerate a cock blackbird among his feet, and sniff tentatively at it when it poked its beak into his oats.

Mallards in the floodwater at the bottom of the field he accepted, but he charged down a crowing cock pheasant and sent it rocketing over the trees. A leveret, which I guessed to be about eight weeks old, arrived in the field, and he accepted it with a whinny and a toss of his head. That leveret did a lot of growing beside him before I moved him to another field to join a mare and a filly — all blacks like himself. Crows he still chased; curlews he did not. Then came the day when he drove a sapling fox to the fence. It misjudged its snake-through and fell almost at his feet. It got through at the second attempt as he lashed out at it with his hindfeet. He was a gey little horse as all who knew him agreed.

Hares on farmland soon get to know the livestock, and how to get along with them, and the *Pax Bos-Lepus* is rarely broken. When there is a happening, a confrontation, it is almost invariably started by the hare, fearing for the safety of small leverets when the big bovine feet come too close to them. I have

Young Carrion Crows just out of the nest.

seen a hare turning away two cross-Hereford cows by her protests, and another driving off a Blackface ewe with her lamb. The pig is a beast that hares rarely become acquainted with. I had a pig called Grumphie, which Jess reared on the bottle because his was a mouth more than his mother had teats for. He used to come for walks with me, and the first time he flushed a hare from her seat he gave chase. The hare ran away but not very far. She must have realised that he wasn't going to catch up with her, and started running rings round him, playing with him until he became exhausted and lost interest in her. I've seen a hare do the same with a fat Sealyham terrier, letting him get close then running away for a short distance before sitting down to give him the come-on sign again. I had to rescue the fatty from himself before she ran him into the ground.

My old friend Fergie Ferguson, retired head stalker, told me of the day his gutsy terrier Sionnach took on a big red deer stag. The dog had run down to the river and wasn't coming back to the recall, so Fergie went hurrying down the steep bank to find out what was happening, and there was a sight for any man's eyes to disbelieve: a big stag in a pool, between rocky banks, stumbling, rearing, swinging his head, with a terrier fastened on the end of his nose. The stag tried everything to get rid of the dog, dipping him, tossing him, and shaking him. Then he tried

Red deer stag on low ground in winter. In general, the open Highland hill is a poor place for red deer and they show this in the small antlers and small body size compared with the giants of Central Europe.

to pin him against the rocks. Fergie could do little to help the dog. In the end the stag held him under for a few seconds, and the ducking made him release his hold. The stag, realising he was free, trotted off. That same terrier was a foxhunter until he became toothless by leaving his top and bottom incisors in the hide of a vixen he couldn't hold. I think these two incidents could safely be called one-offs.

Before Hitler's War there used to be shops which sold nothing but eggs, and one day a poacher friend of mine bought a dozen duck eggs in one such shop, but instead of eating them he put most of them under a broody hen. All the eggs hatched and all the ducklings turned out to be mallards, some of which stayed on and interbred with the domestic ducks the following year. Tame mallard drakes, obviously from this brood, visited neighbouring farms, where they did the same. Wild drakes often associate with domestic ducks and interbreed. It happened with my Khaki Campbell ducks when they went down to the river at Luggiebank. One of my Khaki Campbells snatched up a chaffinch from the ground where it was gathering food and gulped it down before I could catch her.

Domestic poultry on free range — a rare sight nowadays — will sometimes lay away from the henhouse: in hedge bottoms, gorse thickets or haysheds. In the days of stackyards a favourite

Mallard duck.

laying-away place was a hole in a corn stack. The crows used to watch the corn stack nests and when a hen flew out, cackling for one and all to hear, one of them would be down, and in, then out with an egg. At Luggiebank I saw a crow taking eggs to its nest. A Buff Rock hen of mine built up a clutch under a dog-rose, and I noticed that when she came off her nest after laying she didn't cackle *annunciamente* to the world until she was back in the yard. A Blue Andalusian cross-bred hen at a Stirlingshire farm came home one day with a brood of pheasant chicks and neither the farmer nor I could begin to think of an explanation for that. Such was the relationship between laird and tenants in those days (the early 1930s) that my farmer friend immediately handed over his hen and her brood to the keeper, who accepted them as of right, although he had no such right.

Once in a while a stoat or weasel would snatch a poultry chick from a brood running on grass with a hen. One stoat had a cache of eggs under one of my neighbour's henhouses. Another neighbour lost a bantam pullet to a stoat, which didn't repeat the offence. At the time I had a rumbustious Old English Game bantam cock that beaked and spurred a weasel back into the drystane dyke from which it had come. In those days I had no opportunity to teach the cats to leave weasels alone. At

Kathleen with Lisa, and Bounce the young roe deer.

Firknowe I had my tame pair, Tammas and Teen, and had no trouble teaching the cats to leave them alone. It was the same with the terrier: a weasel was a weasel, whether Tammas or Teen or a wild one. Lisa, my German Shepherd bitch, who fostered the roe fawns Bounce and Sanshach and was their friend for life, could never understand why the wild roe ran away from her when all she wanted to do was pass the time of day with them or give them a friendly lick. Her puzzlement was plain to see.

My home ground isn't peregrine country, but from time to time I see one passing through, and keepers shot a few before and after the war. The bird will take homing pigeons whenever they are on offer, and at Palacerigg I had a visit from two owners who wanted to know what they could do about a peregrine that had killed a few pigeons. They knew the bird was protected and didn't want to break the law by shooting it. I told them the bird was on passage and likely wouldn't stay long, and suggested that they should confine all the birds to their lofts for a couple of days. They did so, and by the time they were let out again the falcon had gone.

One of the brightest bird characters I ever had was Pruk the raven, who became greatly attached to Jess, preening her hair, nibbling her ear-rings and sleeping on her shoulder. Yet, with

Tatty howking outside Cumbernauld New Town.

women he didn't know, he was a savage, flying on to their shoulders and going to work on their heads with his powerful beak. I had to confine him to quarters when I had such female visitors. With strange men he was well behaved, contenting himself with stabbing their shoes or tugging at the laces. He used to go out with my neighbour when he was shooting rabbits or crows, riding on a shoulder until the gun was fired; then he would leap on to the barrels, side-step along to the muzzle and try to catch the wisp of smoke coming out. Although he would ride on my Labrador's back he made the farm collie's life a misery, driving it off, time and again, with its tail between its legs.

It has always intrigued me how quickly young mammals, and some birds, will attach themselves to the domestic dog which, in a free-for-all, would be their mortal enemy. It has also intrigued me how quickly the dog, behaving at first because it has been taught, becomes attached to them, acting as their protector and sometimes becoming very possessive. My own dogs have accepted hares, rabbits, weasels, roe fawns, red deer calves, badgers, foxes, hedgehogs, squirrel, polecats, raven, crows, magpies and jackdaws. Polar, our tame greylag goose was, I think, their favourite bird, as the wolves Marquis and Magda were Lisa's favourite people. I use the past tense because Lisa is dead. The wolves, now fourteeen year old, have always been Jess's favourite people.

Lisa at home in her basket with 'her' fox cubs.

Jess with Lisa and Magda, the female wolf, in her enclosure at Palacerigg.

# RIVERS, SEASHORES
# AND ISLANDS

Gannets gliding in the up-draughts of the cliffs of Ailsa Craig.

Colonial seabirds, nesting on cliffs or uninhabited islands, have a high toleration threshold for human beings. On Mull I used to hold speech with Aristotle — a shag, not the philosopher. I called her after the ancient Greek because her kind the world over is known as *Phalacrocorax aristotelis aristotelis* (L) which is not Greek. I always visited her bearing gifts, one of them a small photograph of herself, which she grabbed and tucked under her breast. The shag is a compulsive tucker-under. After accepting a gift she would look at me with her glittering, emerald eyes, and croak about three words in my ear. Besides her photograph she had under her a few hairs from my head, which she had taken on my first visit, some bits of blue wool, a dead and flattened bit of purple loosestrife I had given her, and other bits and pieces of floral decoration which she must have gathered herself.

*opposite*. Green Cormorants or Shags. The name comes from the bottle-green plumage and green eye. They are slimmer, smaller and daintier than the Common Cormorant.

41

Also under her she had three wee gargoyles in brown cotton-wool down, whose style of feeding was to reach half-way down her throat to transfer the fish mix from her gebbie to theirs. I keep calling her female although I had no idea whether she was her or him. I was simply going by the age of the nestlings, and assuming that herself was mothering the nestlings while himself was at the fishing.

One day I was sitting above the beach watching the general scene, pinpointing here and there with the binoculars, when I saw a great black-back gull doing the rounds — scavenging, harrying and pinching. I was used to the sight of herring gulls and grey crows robbing nests, but I hadn't seen the big one around before. So I kept the glasses on him. Half the time I couldn't see what he was doing, but when I saw two oyster-catchers swooping at him I knew what he was at. By the time I reached the nest he had eaten one egg and broken another, leaving the oyster-catchers only one, which they succeeded in hatching.

Back on my seat, which was a great place from which I could see wheatears carrying food to their young, ringed plovers and sandpipers, gulls sitting and oyster-catchers, I put the glasses on Aristotle on her nest. I could see the big black-back in the background, flying low over the shags, which were grunting in anger at him, off their seats but still covering their eggs or chicks. He managed to steal a chick from one bird, and carried it to the tide-line where he tore it up. I went down to see Aristotle, and talked to her for a while. There was not sign of the big gull when I left, but I knew he would be back. He got nothing from Aristotle, whose three small young lived to become big young.

The big black-back is an extremely powerful and rapacious bird, fit to kill a sickly ewe or weakly lamb. It will turn the lamb's skin outside in while it eats the inside out, as it does with birds like puffins, ducks and shearwaters. It eats eggs, including the eggs of other gulls, and it kills young birds, including the young of other gulls. Yet it isn't uncommon to find a single pair living amicably in a colony of gulls of another species. The big gull is bold in defence of its own, and doesn't hesitate to raise the wind round the ears of anyone approaching too close to its nest.

Wind about my ears is the most I have ever had from the big black-back, but I have many times taken direct hits from the lesser black-back, twice to the effusion of blood from the roof of my skull. The fiercest lesser black-back I have ever met was a Lunga bird which came at me from behind while I was looking at her chicks, and furrowed my skull; then, while I was wiping the blood from my eyes, came in again and ploughed a second one. And all that despite my thick thatch.

The great skua, called the bonxie in Shetland, is a robust bird, not noticeably smaller than the big black-back, and dark brown all over except for a white patch at the root of its flight feathers. Like the big gull it is a pirate and a predator, making sustained attacks on terns, gulls and gannets until they are forced to drop what they are carrying or disgorge what they have swallowed. Even the big black-back is sometimes a victim, unable to match the skua's relentless ferocity and aerial manoeuvres. The bonxie is one bird you don't have to go looking for; if you're on bonxie ground the bird will come looking for you.

I went to Noss to meet my bonxies, partly because the idea of having an island to myself for a week or more appealed to me, and partly because Richard Perry had worked there and I was familiar with his work on the bird. I put up two tents inside a sheep fank, and weighted the walls with stones, but despite such precautions I was to lose one of them, ripped from its

A Great Skua at nest on moorland on Unst, Shetland.

moorings and blown down to the sea. Wind was an ever-present problem, and on the high cliffs the gusts were dangerous. But it was exhilarating, and there were always the bonxies.

As soon as the tents were up I went to look for them, and was duly confronted by the first one at the frontier of its territory. I was also duly skelped on the head by several pairs of webbed feet. After finding my first nest I took a dive, stubbing my toe on a peg, which must have been one of Perry's old markers. I felt that the clock had been turned back. On my way off the bonxie ground I found my first wrecked eider nest — egg shells and down scattered round the nesting hollow. The nest was within twenty yards of a bonxie's. The wrecking was bonxie work and I was to see a lot of it during the next eight days.

The bonxies of Noss don't confine their predation to eider eggs and ducklings. I watched them harassing ringed plovers and oyster-catchers. At the Noup of Noss I watched them

Arctic Skua at nest in Shetland. They are slender, streamlined birds with two colour phases, the dark birds being sooty brown all over and the light ones a paler brown above, with whitish neck and underparts.

patrolling the gannet flight-lines, attacking homing birds, driving them right down to, and sometimes into, the sea where they were compelled to disgorge. From time to time I saw thrawn gannets gripped and shaken by the pirates, and sometimes thrown into the sea. Homing shags were not molested during any of my watching spells. Nor was there ever, as far as I could see, an attack on the lambs, which shared eroded rabbit burrows with rabbits and feral cats.

Arctic skuas are slender, streamlined birds, finely drawn in flight, slight when compared with the beefy bonxie. There are two colour phases, the dark birds being sooty brown all over and the light ones a paler brown above with whitish neck and underparts. Apart from colour the Arctic skua can be easily recognised by the two long, straight feathers projecting beyond the rest of its tail. Like the bonxie it is a pirate and a predator. I found this skua very prone to distraction display, what we used to call injury feigning, or the old broken wing trick. Some birds displayed from eggs, and one kept up the act for two days after her nest had been robbed.

One dark bird went through a ritual every time I approached her nest. She would sit on her tail, tilt forward on to her face, lie spreadeagled, tap dance in circles, lurch and stagger and fall on her side. There were times when I became almost convinced that she had acted herself into a kind of trance. Like the bonxie the Arctic skua usually attacks intruders near the nest, and both press them home more and more vigorously as the eggs near hatching. I discovered that both develop a high tolerance of a person remaining motionless for long spells, and in the end I had birds returning to their nests while I was standing less than eight feet away, upright and as obvious as a telephone pole.

A group of Puffins on rock. Puffins often come back to the nest with half-a-dozen fish held across their beaks. How do they catch the sixth without losing the fifth?

There is no beak like a puffin's, so there is no bird like a puffin; without the yellow, red and blue on its beak it would simply be just another auk. I spent most of the day with the Noss puffins, watching them splashing into the sea or homing at high speed, with their catch of fish held crosswise in their beaks, and touching down with legs wide and webs spread. The slope was honeycombed with their burrows, some of which housed rabbits or feral cats instead of puffins, and I soon discovered that the cats were a bigger threat to the birds than the bonxies. In fact, I never saw a bonxie harass a puffin over the sea or standing by a burrow.

The eider ducks on Noss nested in considerable numbers on the breeding grounds of the two skuas — their greatest enemies

A close up of a Puffin. This comical looking bird is distinguished from the Auk by the colour of its beak.

— and this came as a great surprise to me. In some cases skua and eider were sitting within a few yards of each other. Most of the ducks eventually lost their eggs, and those that hatched a brood were harried all the way to the sea, losing most of the ducklings on the way. Yet, despite such heavy predation, I learned that the ducks were actually increasing in numbers at the time.

On some days I saw gannets fishing in small groups, but nothing like the spectacular display I have seen put on by the birds from Ailsa Craig and Sula Sgeir. One day, in particular, remains in my mind. I was sitting with Jim Lockie on a cliff at the most northerly tip of Lewis when a few gannets appeared and began to fish. Presently others began to arrive, in twos and threes, and kept on coming until there must have been a hundred of them, diving headlong, corkscrewing into the water, and fairly bringing the sea to the boil with their splashing. The build-up reminded me of the way gulls see other gulls following the plough and fly to join them, so that the farmer, after only a few furrows, has a great flock following him instead of the half-dozen or so he started off with.

A close up of a Gannet. The formidable beak is used to catch mackerel after a 30 foot plunge.

When I want to get in among gannets I go with my friend Jack Gibson to Ailsa Craig where he has been doing an annual count for forty-three years. For the past years he has counted the nesting birds without going ashore. He has mapped the cliffs off into sections, like pictures on a wall, or a jigsaw, and counts the birds in each section from a boat, using high power binoculars. He does each section twice and is seldom a nest out either way.

Jack is a tireless counter and recorder and has produced check-
lists of just about everything that walks, crawls, creeps, swims
or flies, vertebrate and invertebrate, for the Clyde area. He
knows every hole and corner of Ailsa Craig and it is always a
pleasure to sit among the gannets with him. It angers him that
so many chicks and eggs are lost each year, thrown over the
cliffs during the massed stampede of gannets every time some
fool blasts a ship's hooter to provide a spectacle for passengers.

Guillemots, razorbills (with beaks like cut-throat razors),
puffins, shags and fulmars have a high toleration level for
human beings at nesting time. So has the fulmar, and I have
talked to many a one, at eye-level, from a distance of three feet.
Many people, including myself, have had the experience of
being waxed by a fulmar. When startled or alarmed the bird
will sometimes bock up and squirt a waxy oil in an intruder's
face. It is most unlikely that the bird takes deliberate aim; when
it is face to face with a person the jet almost inevitably goes

Razorbills flying in to the
nesting cliff. They nest in
sheltered crevices in the cliff
face unlike Guillemots
which nest cheek by jowl
on wide, exposed ledges.

from face to face. Probably the act is reflex, not like someone spitting in your eye or throwing an egg at a politician, but more like a partridge going through its broken-wing routine.

Compared with the clamour of the bird cliffs and islands, seashore and estuaries are quiet places, where the birds raise their voices only, or mostly, when alarmed. Round the mainland coast, and the Northern and Western Isles, I have my favourite beaches and estuaries, which I have visited often over a great many years. Sometimes I would be photographing the shore birds; at other times I would do no more than sit down for several hours to watch the passing scene. To sit and stare, or stand and stare like W.H. Davies, at the blue tit on the sloe or the eagle on the height, is for me part of the quality of life. That's what I once told a farmer who had said to me that for a man who got around so much I did an awfy lot o sittin doon. When I told him that having birds to look at was part of the quality of my life he told me that his quality of life was looking

Oystercatchers. These sea birds used to nest on pebbly sea-coasts. Then they spread inland up rivers and now nest on river and loch edges or on agricultural land. They still winter on the sea-coast.

at his bank balance.

One of my favourite beaches is on Loch Striven-side, where I can sit and watch ringed plovers, sandpipers and oyster-catchers, stonechats and wheatears, red-breasted mergansers sailing by with their broods, or the domed head of a seal surfacing like a diver's helmet. The little ringed plovers, whisking this way and that, side-stepping and tap-dancing or tilting forward on unbending legs, are a delight to watch although it is easy to lose track of them because of their camouflage. At one place a sap of water runs from under the access road to the sea, forming the territorial boundary between nesting ringed plovers year after year. One year an oyster-catcher had a nest only four or five yards from a sitting plover. On the other side of the channel two ringed plovers were nesting. If one of the plovers on the oyster-catcher's side crossed the boundary it was driven off by the nearer of the plovers on the other side. If one of the oyster-catchers crossed it received the same treatment, and on one occasion two ringed plovers actually pulled its feathers to speed it on its way. But at low tide the plovers and the oyster-catchers mixed, fed and passed each other without any sign of animosity between high-water mark and low.

One of my favourite estuaries is a wee one, at Aros on Mull,

Young Herons in March, not long hatched.

Heron fishing.

where I have often watched, and filmed, herons fishing the pools and shallows at low tide. The bird stalks thigh-deep, and slowly, in the same way as it stalks voles in a flooded field — half-crouched, with neck in a forward S-bend, and beak pointing down at forty-five degrees. The Aros birds, scooping prey from the surface, dibbling just under it, or striking hard and deep, are often harassed by swooping gulls and hoodie crows, that often succeed in forcing them to take flight and find another fishing place. I don't know why the gulls and hoodies bother because I have never seen a heron disgorge anything for either. Common seals often haul out on the rocks between Aros and Salen and can be watched easily from the road. On the same stretch of road I have live-trapped a Mull polecat, and caught others in the headlamps of my car at night.

The shelduck is a big, goose-like bird, with red bill and pink feet, and plumage of contrasting black, white and chestnut. Drake and duck are look-alikes, except for the front of their faces; the drake has a bulbous nose, or knob, at the base of his bill. The eider drake is another kenspeckle bird; apart from his back crown the rest of him is white above and black below. The red-breasted merganser drake is also easily recognised by his

A pair of Black-headed Gulls on the nest. These gulls lose the chocolate-colour on the head in Winter when the hen becomes white except for a dark spot behind the eye.

Waterhen on nest. Although secretive birds, they are common by streams, ditches, ponds and even in built-up areas.

*opposite*. Heron on nest, in reeds. This is unusual for they prefer tall trees for nesting; but where trees are scarce they will nest on cliffs too.

*opposite*. Grey seals in the sea off a black, rocky, island coastline.

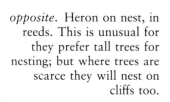

bottle-green head, with its twin-tufted crest, and his conspicuous white collar. One of my favourite places for watching all three was Glendaruel.

As early as the early fifties, when I first fell in love with this little haven, I was surprised by the shelducks' choice of nesting places. This duck likes to nest in holes, especially in rabbit burrows, often as far from the entrance as I am tall, and I stand six feet three inches in my bare feet. In a place where shelducks were common fauna and rabbits legion I expected to find the nests in rabbit burrows near the sea. I didn't find more than one in any year. The others went to the mountain, in and above the forested area, and one year, when I was sitting in a hide watching nesting peregrine falcons, I saw a shelduck, not far below, heading down seawards with her brood. That year there were four families in the sea near Shellfield. At first each was looked after by the parents. But soon the broods could be seen in charge of one pair of adults, while the others were absent doing something else. I had no way of knowing if each adult pair took a turn at baby-sitting. It was much the same with the mergansers. The duck will look after her own family all of the time where the birds are thin on the ground, but where they are common several broods can be seen in charge of a single duck. The eiders are almost exactly the opposite. Once in the water the ducklings attached themselves to the nearest duck and were looked after by her. Growing eiders come in a wide variety of plumage markings and look like a lot of mongrels.

The handsome goosander, a saw-bill like the merganser, is a freshwater species, most usually to be found on rivers and lochs in wooded areas. It prefers to nest in holes in trees, but will also use walkways under rocks or cavities under big boulders. Because it uses the same site year after year, river keepers know where to find it. Its survival, and breeding success, then depend on the kind of person the keeper is. If he is one of these above the law the bird is likely to be shot, or at best has its eggs destroyed. Persecution of the goosander is still a ritual with some keepers, because of its predation on young salmon and trout, although it takes a lot of other species — some the keeper doesn't — like the eel and the pike. Good river habitats are also the haunts of the water ouzel or dipper (once shot because it ate fish ova), the kingfisher a rarity of no importance to fish stocks, and the grey wagtail, and sandpiper.

*opposite*. Young Grey Seal a few days old. The coat of the new born seal is cream-coloured and silky soft and smells faintly of the sea. Soon they become soiled from the dirt churned up in the colony.

# WILDLIFE
# IN THE PAST TENSE

Dragonfly on young bracken in June.

Before human beings came to the Tweed or the Highlands some wildlife species had disappeared, or were disappearing, for one reason or another, climatic change being one. But climatic change has had nothing to do with the disappearance of any species in the past thousand years. Since his advent, man has been the main factor in disrupting or changing habitats, and in killing out wildlife species. He is still doing both. His is the power and not very often the glory.

The giant Irish Elk, the pride of the Pleistocene, had died out before it could meet a human being in Scotland. Climatic change was against its continued survival. It was a loss due to failure to adapt to changing conditions. The red deer survived by the adaptation of reduced size of body and antler.

The lemming and northern rat-vole disappeared in prehistoric times, unable to survive the warmer Atlantic climate. The northern lynx, it is thought, survived until man arrived to push it over the edge into oblivion. And from then on he took over as principal actor in the drama of wildlife.

*opposite*. Female Kestrel. Kestrels characteristically hover when hunting. They specialise in catching voles and small birds.

57

In historic times Scotland has lost the wild ox, the wild boar, the native pony, the brown bear, the reindeer, the elk, the beaver and wolf. Others that joined the ranks of the past tense were the kite, the goshawk, the sea-eagle, the osprey, the rough-legged buzzard and the polecat. Of these the goshawk, the sea-eagle, the osprey and the polecat are back in the present tense. So, too, is the reindeer, after 600 years.

The Romans knew the elk, whose last strongholds were in the far north of Scotland, and there is a reference to it in the Gaelic poem *Bas Dhiarmid*:

> Glen Shee, that glen by my side,
> Where oft is heard the voice of deer and elk.

The brown bear disappeared before the destruction of the Great Forest had really begun, probably in the ninth or tenth century. The Romans certainly knew it, and shipped Caledonian bears to appear in their brutal circuses in Rome. Professor Ritchie has stated that man must have exterminated the bear, as there were no changes in climate or food supply to account for its disappearance.

Man was certainly responsible for the extermination of the beaver. It was hunted for its skin which, in the reign of David I, was subject to an export duty. A small-scale introduction to Bute in the nineteenth century proved that it could survive if left to itself.

The wild boar, greatly prized by the nobility as a sporting animal, seems to have disappeared about the time that James VI took the high road to London. But its half-wild and half-domesticated descendants, bred by the natives, roamed the woods for a long time after that, causing a lot of trouble and damage, some managing to survive until the dawn of the nineteenth century.

The wolf outlasted them all, surviving long enough to become British. People who had heard the last wolves howling could not live long enough to meet the first rabbit, whose coming was not yet. The last record of a man killing a wolf, which may or may not have been the last one, was in 1743 near the river Findhorn. The killer was Macqueen, stalker to The Mackintosh of Mackintosh, and his was a strange story. His long dog cornered the beast, which Macqueen took by the throat and dirked to death.

This beast was 'a black beast thought to be a wolf' and the

fact that it didn't attack Macqueen, or make any attempt to defend itself, suggests that it was. But the tradition that it had killed two children suggests the opposite, that it might have been a wolf-dog or some kind of dog, because it is a well-established fact that non-rabid wolves do not attack human beings. If the Macqueen wolf had been rabid the story would have been different. There was, of course, a last wolf, but it could have died of old age, disease or accident.

The next real killing times began in the nineteenth century with the coming of the Game Ascendancy, when there were exterminations, massacres and decimations, mostly in the interest of sport.

On Glengarry estate, between the years 1837 and 1840, the following were killed: Kestrels, 462; Rough-legged buzzards, 317; Stoats and weasels, 301; Common buzzards, 285; Kites, 275; Martens, 246; Wildcats, 198; Polecats, 106; Peregrine falcons, 98; Merlins, 78; Short-eared owls, 71; Badgers, 67; Goshawks, 63; Hen harriers, 63; Long-eared owls, 35; Hobbies, 11; Red-legged falcons, 7; Montagu's harriers, 5; Gerfalcons, 4; Honey buzzards, 3; Tawny owls, 3.

On the Duchess of Sutherland's estate, between the years 1831 and 1834 the following, among other species, were slaughtered: Hawks and kites, 1155; Martens, polecats and wildcats, 901; Eagles, 224.

Some of the killings were incredible. Even the common dipper, called the water-ouzel in Scotland, was not exempt, bearing witness to the single-minded ruthlessness of Scottish gamekeepers and their masters. Ritchie tells us that, on the Duchess of Sutherland's estate, a reward of 6d. was paid for every dipper killed. The 'vermin' list for Reay, Sutherland, shows 368 dippers killed between 1874 and 1879. Lairds and keepers should have lived in Stupidity Street. But these estates were not special dens of iniquity: they were merely typical of the era. Unlike the sad ruins of crofts in the glens, great chunks of Stupidity Street are still with us, in good working order, and well inhabited.

Of course, gamekeepers today don't kill all the species listed here; they couldn't, even if they wanted to do, because their predecessors made sure some of them wouldn't be around in the twentieth century. They are still in the past tense. It is academic as to whether or not the gamekeepers did all the killing. They dug most of the graves.

# CLAN WEASEL

Tammas, grown up, playing in an ashtray at Palacerigg. The pipe gives an idea of the size of this fierce carnivore.

The weasel is the smallest member of the great family *Mustelidae*, the one most often seen by most people most of the time, and looked upon by many as a kind of pocket Dracula. Cry weasel! and watch grown hominids tremble — not all of them but many of them. I once took a tame bitch weasel into the local watering hole after she had been on TV. All present in the bar had seen the programme, and when one of them asked me where my weasel was I took her from my pocket and put her on the bar counter. And immediately the bar was elbowless in the great desertion of the pints.

One of my captive weasels, the mother of that little bitch, did something that weasels aren't supposed to do. She gave birth to three litters in a year, and reared them all except the pair I took from her at the age of a fortnight and brought up on the bottle. The bitch grew up to clear a bar; the male grew to about the size of a bitch stoat, topping 200 g in weight. Both were reared under optimum conditions. Siblings of this pair weighed 180g (male) and 100g (female) at the age of eight weeks.

Difference in size and weight between the sexes is striking in weasels, males being double the size and weight of bitches. Size in both sexes increases northwards in Britain, and Scottish weasels are bigger and heavier than English ones. The young grow quickly, the bitches matching their dam in size and weight

*opposite*. Young weazels in their nest in a dry-stane dyke, aged 2 to 3 weeks at about the time their eyes open.

61

at the age of seven weeks, by which time the males have outstripped her.

Both sexes are territorial. Territory size varies according to habitat and food supply. In a forestry plantation, with high vole numbers, weasel territories are smaller, and voles are the main prey, often the only prey for long periods. A dog weasel holding ten acres in a young forest would require three or more times that in an area of arable farming, with a lot of open ground. In such places the weasel hunts hedgebottoms, drystane dykes and patches of woodland.

The sexes hold separate territories and live apart after mating, but there is some overlapping, which becomes obvious when one catches bitches in box traps inside the male's boundary. In the case of captive animals a pregnant bitch begins to dominate the male, even to the extent of entering his enclosure and stealing his bedding. My friend Dr James Lockie, formerly of the Nature Conservancy, noticed this change of status long before I got around to breeding weasels.

One bitch weasel that I caught regularly in the same box trap became so attached to it that she made her nest in it and reared her family there. I used to leave a mouse or vole at the entrance for her, which she nosed over, then grappled with, scratching with her hindfeet, before carrying it to her family. After a week or so she would take mice from my hand, so long as I sat perfectly still. Before long her nest was a ball of fur. I would not have made such a relationship anywhere else, teaching a wild animal to trust people. But her ground was in the heart of a big area where all wildlife was protected and I was in charge.

The main prey of the weasel is small rodents — voles and mice, sometimes shrews, rats and rabbits regularly, small birds, and sometimes even small fish like minnows. Bird nesting-boxes are a target, but are easily protected with a backing of perspex or polythene. The weasel, in turn, is preyed upon in varying degree by terriers, foxes, eagles, buzzards, hawks and owls.

Doctor David Jenkins saw a white weasel on Mount Keen in Caithness some years ago, and that is the only record I know of in this country. Farther north on its range the weasel changes like the stoat. I have often wondered if David Jenkins's weasel was an albino.

The stoat is a greater weasel, but the two differ in more than size. The stoat has a bushy tail, from four inches to five and a half inches in length, which is tipped with black at all times.

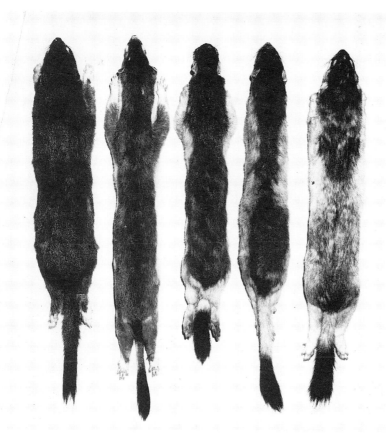

Museum stoat skins showing in the manner of change from summer to Winter coat. In Britain the change is not always complete, and piebald stoats are common.

The line of demarcation between the brown upper fur and the lower white is straight whereas in the weasel it is like an oak leaf. The stoat doesn't have the two brown spots under the chin which are found in all the weasels. On average the dog stoat is half as big again as the bitch.

In its white winter coat the stoat becomes the ermine of commerce. Stoats in the north moult to white every winter. Farther south there can be three phases at any one time at the same latitude — a moult to ermine, a partial change to brown and white, and a moult to a denser and paler brown. I have seen white and brown-and-white stoats on the same ground in January.

The autumn moult is completed quickly. My pair of captive stoats, in their first winter, were brown when I left home on a Friday evening and white when I returned on the Monday night. The following winter they made a complete change to white, but in the third winter the change was partial. When they were moulting back to brown in spring, there was a period when the

A Stoat becomes an Ermine in winter. The diagnostic black tip to the tail is always present.

face of one was brown while the other's was white.

An ermine on snow is not easily spotted until the black tail-tip begins to move; an ermine on green or dark ground is as easy to see as the moon in a cloudless sky.

Unlike the weasel the stoat has only one family in the year. Mating takes place in June/July and the young are born in April of the following year. This seemingly long pregnancy is due to delayed implantation, the blastocysts lying loose in the uterus until the following March when real development begins. The young are born about a month later. Delayed implantation explains what was a great puzzle to naturalists of old — how animals like stoats and badgers, kept captive for many months, could suddenly produce a family without seeming benefit of male.

Litter size varies, from six to twelve by the record. I've seen litters of seven and eight, and watched a family party of eleven one harvest time playing follow my leader on a farm wagon in a stooked cornfield. The females of the litter mate soon after weaning; the males are not ready to breed until the following spring. The dog stoat, unlike the dog weasel, helps in the rearing of the family, which stays together even into November, to become the autumn pack so often seen and reported in days when stoats were much commoner than they are now.

The so-called winter pack is something else, and I have seen five such in 60 years, always during severe winter weather of lasting snow and bitter frost, and the beasts were, to my mind, shifting ground because of hunger. One group of sixteen, presumably two families — perhaps a mix of clans — flowed literally round my feet without taking the slightest notice of me. The only time I ever had an argument with stoats was in 1934 when I tried to take a rabbit from a family party and was bitten twice after I had stumbled to my knees.

The stoat is a much shyer and less demonstrative animal than the short-fused weasel, although it can make all the same noises, including the angry ones and the croodling of bitch to young or mate to mate. Catch up a wild adult weasel, and you will find you have a bold, truculent captive that will stare you in the face next day, and maybe shriek at you, and on the next day will take a mouse from your hand. Do the same with a wild adult stoat and you will be lucky if you see it, let alone hear from it. Young born in captivity grow up to be as extrovert as the weasel.

Stoats are as playful as otters. I've watched a bitch spinning on her seat on top of a straining post as though it were on a turntable. Then she did a sloth crawl along the top wire, dropped to the ground, and ran round and round the straining post. I squeaked to her and she sat up tall to stare at me, not recognising me for what I was. When I backed off she was on top of the post repeating her turntable act.

Before myxomatosis the rabbit was the stoat's staple diet and it took all sizes, hunting mainly by scent and tracking the selected prey to the kill, even when other rabbits were feeding closer to hand. Depending on the size of the rabbit the stoat will carry it away, drag it away, or feed from it on the spot. No stoat can move a fully grown rabbit. My own pair failed with a big one and I filmed their failure.

Nowadays the stoat has to be more catholic in its tastes, and it would be true to say it takes anything it can catch and hold, and eat it if it is palatable. It kills rats, mice, voles, small squirrels, leverets, birds and insects. It climbs well and my wife and I watched one kill two woodpigeon squeakers in their nest. It also swims well so it can reach the nests of coot and moorhen. It will sometimes turn its playfulness into play-acting, cavorting about until it attracts a curious audience of small birds. If its cavorting takes it close enough to strike it will pounce and sometimes catch. Its main enemies in this country are gamekeepers and those who shoot for 'sport', which pleases them but does little of significance to help their own prey species.

The polecat (foul marten or foumart because of its stink glands) is a size up on the stoat, like a big polecat-ferret. Polecat skins were once exported in thousands from Scotland. Then, in the nineteenth century, came the Game Ascendancy, and the polecat was shot, trapped and poisoned out of existence. Shortly after the turn of the century it was extinct, a salutary example of man's power of destruction for selfish ends. The polecat lasted a lot longer in England, holding out in two strongholds

David, David Steven's grandson, getting acquainted with one of the polecats.

until 1930 or thereabouts. But it survived in Wales, whence emigrants are now colonising the English border.

What about Scotland today? There are now polecats in several parts of the country, mostly Welsh or the descendants of Welsh, released into the wild by Anons who aren't telling the world where. I have farmed out young polecats bred by myself to be bred from by other people. There is nothing wrong about reintroducing a lost species. We've done it with the capercaillie, the sea-eagle, and the reindeer.

The polecats of Mull are polecat-ferrets, descendants of their kind introduced there a long time ago to help keep down rabbits. I've seen a lot of them there, and over a number of years, using a succession of imports, have bred them at home. The young came in a variety of shades and markings, from the whites used mostly by ferreters to animals so like the real thing

Polecat in dyke. These animals have been re-introduced here and there in Scotland but there are also many escaped polecat ferrets in the wild. Ecologically they are identical to polecats and often look very similar.

that they fooled a lot of people. I also crossed the pure polecat with the best of the Mull types and bred a number of very real looking specimens. I got around to referring to the true polecat as the Right polecats and the Mull ones as Wrong polecats.

But how does one tell the Right from the Wrong if they are look-alikes? It can be done, it is said, by examination of the skull. It can't be done by looking at a living animal that appears to be the real thing, and may or may not be. Researchers Ashton and Thomas say that the only difference in the skull is that the post-orbital constriction has a 'waist' in the ferret but not in the polecat. But another researcher, Tetley, discovered that the few skulls of the Scottish polecat still in existence have this 'waist'.

In my book *Highland Animals* (1974) I summed up my own thoughts on Right and Wrong polecats thus: 'It seems to me that a polecat that looks like a polecat, acts like a polecat, is accepted by other polecats as a polecat, and breeds polecats that look like polecats, is a polecat for all practical purposes. The matter becomes academic if the beast happens to raid your hen-house. Real or ersatz the result is the same.'

A big dog polecat can weigh up to two and three-quarter pounds, and reach a length of two feet, about seven inches of which is tail. Bitches are smaller and lighter. An animal of this size can kill big prey, which in Wales is mostly hares and rabbits. Voles and mice are also taken, and frogs are a common prey. The polecat also kills hedgehogs and poultry. Like the stoat and weasel it kills by a neck bite, and will lick blood spilt over. None of them sucks blood.

The pine marten, the graceful forest weasel, is a size up on the polecat, the only animal with which it is likely to be confused. It is shy and secretive and, until recent years, seldom seen, and we used to talk of sightings rather than observations. More and more, with its own increase in numbers and colonisation of new ground, it is being seen by more and more people, especially in car headlamps at night. But it is also being seen by more and more people without benefit of headlamps, and I have letters from people who have martens coming to their bird tables in daylight. Another correspondent told me of a pair in his loft that went rampaging around and made ever increasing demands on him for bread and jam.

Sightings, while they didn't tell anyone very much about pine martens, at least indicated when they were most active. James

Young Buzzard just out of nest in mid-June, its tail feathers still sprouting.

Lockie, who did a lot of research on this beautiful weasel, found that, in twenty sightings, four were at night, five during the day, and eleven around sunrise and sunset. This tallies with my own fewer sightings, but oddly enough my longest one was by day, in 1951, at the height of that severe winter.

I was motor-cycling in the north-west on the morning after a blizzard. Buzzards were down on the road and grouse were running ahead of me like refugees. I was paddling the motor-cycle round a drift when the scolding of small birds in some scattered trees attracted my attention, and there in one of them was a big pine marten, a giant weasel with a fox-tail and a pale lemon throat patch. I stopped the motor-cycle and watched it leap from branch to branch like a squirrel. Then it came down the trunk backwards and ran to another tree, bounding with rump arched. I laid the bike down and followed on foot. When it left the second tree it bounded uphill and out of sight, leaving tracks as like a hare's as could be, without being a hare's.

A tree weasel the marten is, but it spends much of its time on or near the ground, where it does most of its hunting. Lockie did some research on the food of martens in the Scottish Highlands and found that rodents and small birds were the main prey in all seasons: field voles, tits, wrens and tree-creepers. Other items were carrion, beetles, caterpillars, birds' eggs, fish and berries. My own captive marten ate rabbit, vole, woodmouse, frog, venison, poultry, caterpillars, beetles, eggs, squirrel and fish. He was very fond of rowans, raw or cooked to a jelly. H. G. Hurrell's tame martens, when allowed to hunt, killed rabbits, rats, frogs, bumble bees and cockchafers. The marten is a squirrel hunter — in Finland and Sweden squirrels are a staple — but apparently not so in Lockie's study area, where they were thin on the ground and not readily available.

Although the marten is a forest animal it proved that it could survive on the most rugged terrain when it was persecuted almost out of existence. Keepers and game preservers killed it as a matter of routine. One of the greatest butchers ever let loose in the Highlands was Charles St John who, after describing the beauty of the beast, told how he shot it. He shot every marten he came across, and when he was not shooting them he was cheering dogs on to worry them. There were many like him. In 1920 Dr Ritchie was warning that the marten was an endangered species. During, and after, the Second World War it increased its numbers and spread into new country, helped to some extent by reafforestation. It would have benefited even more from mixed forest rather than solid stands of conifers.

It is difficult to envisage a Scotland in which 245 pine martens could be killed in one estate (Glengarry) in the four years up to 1840. A hundred years later it was still a rare animal. Why was it butchered so ruthlessly? Because the high priests of the Game Ascendancy ordained it so. The marten killed game birds and ate their eggs, which everyone knows is the Eighth Deadly Sin. It is now a protected species, for what that is worth.

The marten breeds once a year. Mating takes place in late summer, and there is delayed implantation until the turn of the year. Three young are usual and they are fully grown by summer of the following year.

Once upon a time the otter was found over most of the country, except in built-up or industrialised areas. Today the picture is different. For a number of reasons, including pollution, disturbance of habitats and river-control structures, it has suffered a disastrous decline, and it is now protected by law. Its main strongholds are now the Highlands and Islands.

In many ways the otter resembles a seal. It is streamlined, has powerful nose and ear muscles that snap shut before a dive, and a heavy bush of sensitive whiskers set in the swollen upper lip. It is, in fact, a web-footed weasel, supremely adapted for its dual role of land- and water-hunter.

Although the otter's main business is catching fish, anything from minnow size to big game fish, it is also an opportunist predator, taking what presents itself: rabbits, rats, water-voles, field-voles, frogs and newts. Coastal otters, mainland or island, prey heavily on crabs, other crustaceans, mussels, and probably saithe, lythe and other rock fish. When hard pressed the beast will eat gulls and their eggs, poultry, ducks, leverets and carrion.

I once watched a bitch otter with a single cub catching frogs in an old quarry pond at spawning time. They caught them by pushing their heads under water or snatching them up from the shallows. The pond was muddy and the otters became plastered on head and neck, but a turn in the water soon cleaned them up again. One moonlit night, during the savage winter of 1946/7 I watched an otter killing a duck on a frozen loch. The otter came on to the ice via the bank shadow cast by the moon, stalked through the shadow, launched out at great speed, and caught a duck, while the others beat up into the face of the moon.

A long time ago I lived beside the Luggie Water which was a favourite for travelling otters, in both directions. One year I freed a big male from a gin set by the foxhunter. A few years later a dog otter got off course, probably dogged off, and was set upon by two greyhounds. He gave them a running fight until he was killed by their owner with a blow on the head. He was a fine beast, weighing twenty-three pounds, and I gave the body to the local museum.

Otters are territorial, and according to *ci devant* hunters, were spread out evenly along the better river habitats at intervals of about five miles. Lockie reckoned this about right for Wester Ross and the Western Isles. On Lewis I reckoned it under that. In Sweden the researcher Erlinge calculated adult male territories at diameters around fifteen km and about half that for a bitch with cubs. Otters deposit their spraints (dung) at a number of places on the territory, usually on a mound or tussock or a molehill where they can be easily seen. If there is no hummock the otter will sometimes scrape one up.

The breeding holt of the otter is usually away from the main stream, and as often as not has an entrance under water; others are heavily screened by washed-out roots or other vegetation. Some breeding nests are in woodland, or thickets, and Lockie found one in a dense clump of rhododendrons. Rock holes and sea caves are used by marine otters.

There is no delayed implantation in the otter, and practically no evidence indicating a peak birthing period in spring. Births have been recorded in all months. Litter size varies from two to five, but two or three cubs are usual. Cubs begin to swim when two or three months old, helped and closely tended by the bitch. The dog otter's role as father of a family appears to be no role after the cubs are in the water. Yet I once watched two adult otters playing with two cubs near Aros in Mull. Dog and bitch?

It used to be argued that otters reduced the stocks of game fish. This charge has never been demonstrated, and the argument is now academic because the otter is the one in decline. It was called wasteful because it sometimes ate only a bit of the shoulder of a big fish and left the rest. Too much has always been made of this. Like other members of the Clan Weasel the otter can be choosey when food is abundant. At other times it will eat every scrap.

The home of the American mink is Canada and the United States. It has been ranched in Britain on an increasing scale since 1929, and there have been escapes almost from the beginning. Today's feral population is widespread, and it would be easier to say where mink are than where they are not. I have caught them in many places, including Palacerigg Country Park, where I put them on exhibition behind bars.

On average the mink is about the size of a polecat-ferret. Ranched mink were bred in several colours, but the commonest colour of the feral population is rich dark brown, with a white spot on the chin and lower lip. Official policy was to exterminate the mink, a hopeless one from the start because ranched animals kept on escaping.

When a feral mink raids a poultry house, or gets in among ornamental waterfowl, we hear about the raid because it usually ends up in a massacre. Its actual prey list must be as varied as in the United States where the bulk of its winter food is aquatic mammals. At other times it takes fish and birds as well. In this country I have watched one catching frogs, and another exploring rabbit burrows. One day I got a report that a mink had been seen going ashore on an islet where one of my Canada geese was nesting. By the time I reached the lochan the mink was in the water again being stabbed at and wing-buffeted by the angry gander. The mink escaped into cover seemingly uninjured.

The heavyweight of the Mustelids is the badger. Boars are a little heavier than sows, on average twenty-two pounds to twenty-five pounds, but both sexes have been recorded much heavier than this in the Highlands, up to forty pounds and more. Most badgers appear grey because the hairs have a pale base with a black band towards the pale tip; the legs and underside are black. In 1962 I photographed a buff-coloured badger. Albinos are seen occasionally. Ernest Neal has described graduations in colour from black to red-brown on the dark areas and through white to cream and sandy yellow on the light.

Mink in winter. These animals are an introduced species, having escaped from mink farms over the years. They are now part of the British fauna, occupying a place between otter and polecat.

Badgers are squat and powerful, bear-like rather than weasel-like, with short, strong legs and long foreclaws for digging. They like rolling, hillocky country, wooded knolls, hanging woods, old quarries, rabbit warrens and coastal cliffs, and can be found in mountain cairns up to, and even over, 1500 ft. They shun marshland and low-lying ground that is liable to flooding. Mountain badgers sometimes lie out among rocks or in stands of bracken.

Although a carnivore the badger is really omnivorous, and eats a wide variety of animal and vegetable foods. Earthworms are one of the main items and it is a fact that tame animals will gorge on them at any time. One of my tame ones, running free at night, was killed near the house. Her stomach was crammed with earthworms. Baits of peanuts or honey will attract wild badgers to a viewing place, and will hold emerging animals near the sett for several minutes. Wild fruits are eaten in late summer and autumn, when the animals also wreck wasp nests for the grubs. They eat acorns and dig for wild hyacinth bulbs, pignuts and other underground storage organs. Animal food includes rabbits and rats, voles and mice, frogs and newts. In the days of free-ranging poultry a badger here and there would occasionally break into a henhouse. Now that poultry are mostly in battery cages or broiler houses a badger probably wouldn't recognise a hen if it saw one.

Mating takes place in spring soon after the birth of the cubs, and if the sow doesn't settle then she will mate again. There is delayed implantation, and most cubs are born in late February into March. They begin to come out of the sett about the age of eight weeks, and are weaned at three months. Before they fend for themselves the sow regurgitates for them, and leaves scent markers as they explore farther and farther from home. Badgers set scent on cubs and each other as well as on the ground.

My own tame badgers rubbed their seat on my boots every time I went into their enclosure; they also set scent on my terrier. When I took them afield they set scent frequently on me and the ground, and on each other if they had been separated for any length of time. When I was living beside the Luggie Water an English veterinary surgeon visited me during his honeymooning in Scotland. With him he had his pair of badgers! The female was named Marylin. She and her mate were put in my stable, from which they escaped when some children opened the door for a peek and didn't close it securely again.

My friend wasn't greatly worried, as the two were great homers and he was sure they would return. When the police phoned to say there was a badger travelling north on the pavement we guessed it had to be one of the pair. My friend went to the stable and got into a sleeping bag to wait. Marylin returned from four miles away, and bored into the bag. Her mate was killed on the railway line less than half a mile away. How had Marylin found her way back? By back-tracking herself via her scent markers.

Many years ago I housed a tame pair in an artificial sett I had built in the wood, just outside the garden wall. They hunted afield every night, sometimes visiting the house first for a titbit, sometimes scratching at the door before daylight and setting the dogs barking. They always liked a romp with the dogs. When the boar was killed by a workers' bus on the road not far from the house the female quit the artificial sett and went back to the wild. But we had visits from her from time to time until the spring.

Despite being protected by law the badger is still subjected to all kinds of cruelty. Baiting is illegal but still goes on. Digging is illegal unless for the purpose of moving the beasts from a place where they are not wanted to a place where they are. Gassing is illegal but still goes on. Sending a terrier to ground to face a waiting badger is a good way of getting it maimed, sometimes to the extent of its losing half its face, and those who arrange such confrontations should resign from the human race.

Young badger at the entrance to its sett.

# THE LORDLY ONES

Young Hen Harriers in heather moorland. The eggs hatch in the order of laying and according to the intervals between them, so the first chick to hatch can be 10 days ahead of the last.

In days of old when knights were bold and falconry the sport of kings, princes, nobles and lesser breeds, the raptors were the lordly ones among birds, their status that of the person who carried them. The place of the goshawk and peregrine was on the fist of kings and nobility. Lowlier species, such as the sparrowhawk and kestrel, were for lowlier people. The lady's hawk was the dashing little merlin, which is a falcon.

How beautiful they were, the lordly ones, and how high their standing in days when a pair of goshawks might have cost £1000, or in earlier times when twelve Greenland falcons could ransom a Crusader prince from the Saracens. With the coming of the gun their status gradually declined; with the improvement in firearms and the introduction of shot they became redundant. In the nineteenth century they were relegated to the status of 'vermin', and their new place became the gamekeeper's gibbet. Their days of glory were in the distant past; their present the constant threat of death by gunshot, trap, snare or poison.

*opposite.* Female Sparrowhawk on nest in Larchwood. The cock, who does all the hunting in the early days, plucks the prey before passing it to its mate, who then carries it to the nest in her feet.

Osprey and sea-eagle are with us again, the first back by its own choice, the second reintroduced by the Nature Conservancy Council. With the passing of the 1954 Act all birds of prey, except one, became protected by law. That sad exception one was the sparrowhawk, which remained 'vermin' until organochlorine pesticides so threatened its existence that it was given special protection. Another pesticide, dieldrin, used in sheep dips, posed a similar threat to the golden eagle in the north-west and the isles. Both were withdrawn, and eagle and sparrowhawk are back to strength.

It remains to be seen how the sea-eagle settles in after its absence of seventy years, and whether it comes into collision with sheepmen again. The old stock became extinct in 1918. The last successful breeding was in 1908, after which the male was killed. The female, an albino, sat on the nest every year until 1918 when she disappeared. Although wandering sea-eagles have been seen from time to time in the Outer Hebrides, there was no known nesting until the Nature Conservancy Council's successful reintroduction on Rhum.

Golden eagles and sea-eagles are much of a size, but the sea-eagle is a much stockier bird, with a more massive beak and a white tail when adult. Immature birds have brown tails. Young golden eagles have a lot of white on theirs, so might easily be mistaken for sea-eagles. The tail of the adult golden eagle is brown or tawny, longer and more wedge-shaped than that of the sea-eagle. The legs of the golden eagle are feathered down to the toes; those of the sea-eagle are not. In both species the females are generally bigger than the males.

Despite protection the golden eagle still takes a hammering from many gamekeepers, and sheepmen, who have other ways of killing than by shooting, trapping or poisoning — keeping the hen off her nest until the eggs are chilled beyond warming back to life; pinpricking them so that they will never hatch; dropping a heavy stone on small eaglets to crush them to death; or cutting their throats.

Before myxomatosis almost annihilated the rabbit population rabbits were a common prey of eagles, and remained so in places where the disease did not strike. A pair of eagles in Kintyre reared twin eaglets almost entirely on rabbits. Unfortunately, someone cut their throats when they were about ready to fly.

At a coastal eyrie on Lewis I expected to find seabirds as

Young Golden Eagle chick about one month old. Eagles are capable of killing lambs but prefer rabbits, hares, ptarmigan and grouse. Many lambs that find their way to an eyrie have died from other causes and are picked up by the eagle as carrion.

prey, and I did. But I also found that the eagles were killing a great number of rats and rabbits. On Mull I kept watch on an eyrie for a season, and found that the birds were killing hoodie crows and rabbits. A pair in the central Highlands, with twin eaglets, killed on average a mountain hare a day during the rearing period of eleven and a half weeks.

In some places, where sheep and lamb mortality are high, eagles live more like vultures, feeding on carrion mutton and venison. On Lewis I've watched two eagles feeding on a dead sheep a hundred yards from the road. I have never seen an eagle strike a lamb; nor have I ever seen one brought to an eyrie. I have handled only one dead lamb that had been killed by an eagle, and stitched up another that had been attacked.

Some years ago a farmer friend of mine, in Argyll, phoned me to say that there were two lambs in the eyrie on his ground. Would I come down and take a look? I did, and took my old friend Jim Lockie with me. The lambs were taken from the eyrie and examined. Neither had been killed by an eagle. Both had died of pulpey kidney and been taken dead. The presence of lambs in an eyrie is no proof that the eagle killed them. I once found poisoned rabbits near an eyrie where the eagles had only carrion.

The following are the prey lists noted by me at four eyries, all on sheep ground, in widely separated places:

Black-faced lamb. This breed of sheep is the principal one in the Scottish Highlands.

(1) Perthshire. Heavy stock of lambs right up to the corrie of the eyrie. Period 12 days. Twenty-three red grouse, fourteen mountain hares, nine rabbits, one red deer calf, one golden plover.

(2) Perthshire. Moderate stock of lambs. Period thirty-one days. Sixty-three red grouse, twenty-seven rabbits, nineteen hares, three hoodie crows, two wood pigeons, one rat, one fox cub.

(3) Argyll. Optimum stock of ewes and lambs. Period ten days. Twenty-nine rabbits, nine hares, three red grouse, two hoodie crows, one oyster-catcher.

Two cubs at the entrance to their den. Fox cubs are easy to rear and become tame quickly, but they should be left in the wild unless one has a good reason to take them from it.

(4) Argyll. Low stock of sheep and lambs. Reafforestation on a big scale. Period seven days. Fifteen rabbits, ten hares.

A pair of eagles on Atholl killed three well-grown fox cubs. A pair north of the Great Glen had seven weasels in the nest, untouched. A Glendaruel pair had a cuckoo. I got a big surprise when I found two adult pheasants in a coastal eyrie on Mull. A pair in the Loch Lomond area had a meadow pipit in the nest one day and a wheatear the next, yet the hen eagle allowed small birds to gather food in the eyrie behind her back. That pair had one fox cub, and I'm sure eagles kill more of them than is generally realised. Indeed, one stalker told me that his eagles were his best fox-hunters.

Following a complaint about lamb-killing by eagles. I went to Lewis in 1955 with Jim Lockie to help him investigate the problem. There were ten eyries in the study area. Few produced anything, the eggs or chicks being destroyed. One coastal eyrie did survive and the single eaglet got off safely. That eyrie was being guarded by the shepherd on the ground! He told me he had never lost a lamb to an eagle in his life, and that he considered the whole business of eagles killing lambs was highly exaggerated.

The buzzard is a lesser eagle, of the same clan as the eagle but not the same sept, its length twenty–twenty-two inches compared with the eagle's thirty–thirty-six inches. Although a far less powerful bird than the eagle, the buzzard has, at one time or another, been recorded taking just about everything on the eagle's prey list except red deer calves and roe deer fawns. But its main prey is small mammals: rabbits, voles and leverets. It is a considerable carrion eater, and will feed on dead lambs, ewe placentae and deer grallochs. It is also a great hunter of earthworms, and I have watched a pair gorging on them on Mull.

Prey varies, but not greatly, from place to place, although there is always opportunist predation to provide surprises. A Glendaruel pair I was photographing had in a weasel and a slow-worm on the same day. A pair on the island of Eigg had in nestlings of small birds, not all of them dead on arrival. I watched the cock bird working a slope on foot searching for nests, the only time I have seen such behaviour. Bird prey is usually taken by surprise rather than pursuit, although I did watch one on Mull running with ouspread wings after a young peewit. The chick escaped into some rocks.

Like the eagle, the cock buzzard frequently brings fresh greenery to the nest during the rearing of the young, and the watcher can never be sure if he will arrive with a prey of a spray of rowan or aspen or whatever. Again like the eagle, the buzzard will feed a dead chick to the living at a time of food shortage.

When soaring or gliding at a great height the buzzard might easily be mistaken for an eagle, because size can't be easily judged so far away. If the suspected eagle mews it is a buzzard. Looked at from below the eagle has an obvious neck while the buzzard appears to have hardly any.

Belonging to a different sept of the clan again is the peregrine falcon. Along with the goshawk it is in greatest demand among present day falconers. Falconers of old, not forgetting William Shakespeare, almost always referred to the peregrine as she, because the female was the one most in demand. Bigger and more powerful than the male she was the falcon, while the lesser bird was the Tiercel. The names still stand among falconers today.

There has been a great revival in falconry, and the number of falconers keeps increasing but it is extremely doubtful if it will ever become 'big time' again. The revival has brought its problems. Young peregrines are taken illegally from the nest and sold for big money. The goshawk is also in great demand. Some smaller birds such as the kestrel and sparrowhawk have been taken from the nest by novices with resultant suffering and tragedy. The film *Kes* has a lot to answer for in this respect. For two years afterwards I had to take a number of kestrels and sparrowhawks into care, injured or weak after escaping.

The peregrine is a powerful bird and a brilliant flyer, a feathered missile when it comes down in the grand stoop to knock grouse, or blackcock or whatever out of the sky. One pair I worked with in Argyll provided a flying spectacle within a few days of each other, the first by herself, the falcon. Some hoodie crows were loitering and gabbing small-talk above the corrie when she came in her power dive, streamlined, and a crow went spinning to earth, a twisted wreckage of feathers, to be ignored after it hit the ground among some rocks.

The tiercel gave me his show when he came back down on a cuckoo that had been calling in a tree on a slope below the corrie. The bird was flying from its calling tree to another some distance along the slope when the tiercel struck. The bird crumpled, and fell, but the tiercel followed it down, took it before it hit the ground, then carried it to his plucking place before bringing it to the nest where his mate was tending two chicks. My friend James Robertson Justice, who was a noted falconer, told me later that he regularly found cuckoos in the peregrine nests he knew of.

I had a hide beside a wheatear nest on the slope below the corrie which I sometimes used to watch the falcons because it gave me a wider view than the one I had built near their nest on the cliff. Grouse were thin on the ground, but one day herself came hurtling along above a lone one flying down on a slant for the sanctuary of the heather. It crashed in only yards ahead of the stooping falcon. Shrieking, the falcon threw up, and around, and down again, then alighted and walked round the perimeter of the sanctuary, high-stepping and flat-footed, lowering her head at intervals to peer through the woody heather stems, seeking the lost prey. She didn't succeed in flushing it. I had watched all this in amazement, because the peregrine isn't given to snow-shoeing around on the ground, although it will sometimes catch a prey there, like a small hare, or rabbit, or a duckling not yet able to fly.

Although the peregrine tiercel takes short spells of sitting on the eggs during the day, his role after the young hatch out is that of provider, while the falcon feeds and broods them. Very often he will call his mate off the nest and pass the prey to her in the air, foot to foot, or drop it for her to catch. At other times he will fly to the nest with it and drop it on the edge, or present it to his mate in his beak. In about three weeks the falcon begins to hunt as well and both drop prey in the nest for the young birds to tear up for themselves.

Best known of all our falcons is the kestrel, which is so common in the countryside and no stranger in open spaces in towns and cities, advertising itself by its style of hunting, which is to hover in full view at a height of twenty–fifty feet while scanning the ground for vole, mouse or shrew. The bird changes station at short intervals, forward or sideways, quartering the ground like an aerial setter, then swooping down when it sights a prey, which it doesn't always catch. The bird is a familiar sight on road-side telephone poles, either taking time off or with a prey under a foot.

Female Kestrel. The Kestrel is often described as a 'useful' bird which means that it is generally inoffensive to man or beast.

More than fifty years ago W. E. Collinge examined eighty kestrel stomachs and found that sixty-five and a half per cent of the food was small mammals, sixteen and a half per cent insects, eight and a half per cent small birds, six per cent nestlings, two and a half per cent earthworms, and one per cent frogs. Between 1946 and 1978 this was still right for my neighbourhood in terms of prey brought to the several nests I watched regularly for a week at a time. But there is some variation with locality. On open deer forest I have found the kestrels taking a slightly higher percentage of birds, mainly the chicks of waders such as curlew, peewit, golden plover and redshank.

David Stephen examining a prey brought to a single eaglet in a Golden Eagle's eyrie in Argyll.

The kestrel is often described as a 'useful' bird, which means that it does no harm to anybody's interests. Nowadays it is shot only by gamekeepers and some people newly come to 'sport' who seem to get the anti-vermin serum with their receipt for the gun. Businessmen phone me from time to time to say that there's a kestrel nesting on, or in, their factory and asking what they can do to help.

The dashing little merlin is smaller than the kestrel, but not all that much, and its prey list about the reverse: mostly small birds instead of mostly small mammals. It isn't liked on grouse moors although, as anyone who has watched nesting merlins at close quarters knows, the chicks would hardly survive the absorption of their yolk sacs if they were to depend on the parents feeding them grouse cheepers. Yet keepers kill them just the same, or give their eggs the order of the boot, that is by trampling them into the ground. A merlin in a tree, using the old nest of a crow, isn't any safer. A shot through the bottom of the nest destroys the eggs, or chicks as the case may be.

The merlin is a specially protected bird, much rarer than it used to be, although not yet an endangered species. In my own neighbourhood I used to be able to find two pairs nesting each year. Afforestation and peat cutting, together with the spread of Cumbernauld New Town, destroyed the habitats. Only game-keepers and shooting men kill merlins.

Another bird killed on grouse moors is the protected hen-harrier, a big hawk, a two-times sparrowhawk, whose eggs are also given the order of the boot and taken to sell to collectors. Before the war the hen-harrier was a rare bird on the mainland, being confined to Orkney and some of the Outer Isles. Anywhere

Hen Harrier female in young pine plantation. The early stages of forest plantation provide a perfect breeding place for these birds.

else it was news. A pair tried to nest in Perthshire in 1922 (and you can read into that what you like) and a pair did breed in Inverness in 1936. After the Second World War it began its spectacular spread, helped by a number of factors, including protection since 1954, the sanctuaries provided by the young plantations of the Forestry Commission, and the goodwill of that body.

The harrier lays from four to six eggs, usually on alternate days but sometimes at even longer intervals, so as she begins to sit with the second or third egg, the chicks don't all hatch at the same time. They hatch in the order of laying and according to the intervals between them, so that the first chick out can be ten days ahead of the last one to hatch. This difference in age is reflected in difference in size, and strength, and puts the smaller at a disadvantage in the hurly burly of growing up if there is a shortage of food. If food is plentiful even the smallest chick can grow into a big harrier.

At first the cock does all the hunting, calling the hen from the nest when he has a prey, and passing it to her foot to foot in the air or dropping it for her to catch. Only the hen feeds the young. Food shortage at a nest isn't necessarily the result of shortage of prey species in the cock's hunting area; he could be a bigamist or a polygamist, with too many mouths to feed.

Hen Harrier chicks tussling over Meadow Pippit prey. The last born often stays small or dies if the male doesn't bring enough food to the nest.

Where there is an assured food supply the young of a monogamous male have a high survival rate. Food shortage can mean the death of the smallest chick, and perhaps also the one above it in the peck order. The hen will feed the dead to the living in such cases. One hen I was photographing carried a dead chick from the nest and I though she was going to drop it somewhere; instead, she came back with it and dropped it in the nest as prey.

Cock and hen are easily told apart, even at a distance, he being pale blue-grey and she brown, with a grey rump patch. Seen against dark spruces the cock can appear quite silvery, to the extent of being mistaken sometimes for a gull. They hunt in slow-flapping and gliding flight, close to the ground, which is where they catch almost all of their prey, although occasionally they will catch a bird by pursuit.

The hen makes a noisy demonstration when her home ground is invaded by a person, and will fly out to meet the intruder, who is quite likely to be assaulted if he or she approaches the nest. She will swoop down to head height sometimes lifting clear at the last moment, sometimes coming in with her undercarriage down and her talons spread. And she will draw blood if you give her a target. The answer is to cover your head and turn your back on her.

Rarity is no Safe Conduct pass for a bird with a hooked beak appearing on a grouse moor, as I discovered when I was photographing the first Montagu's harriers ever to nest in Scotland. The two chicks were about a fortnight old when the hen was shot. The cock began to bring food right to the nest but didn't feed the chicks, and I made provisional arrangements with the then Nature Conservancy to take them into the care. But the cock suddenly took over hen's work. He came to the nest with a bird in his feet and transferred it to his beak. There followed a tug of war between him and the chicks, during which the bird came apart. He then began to feed them small pieces. Later in the day he brooded them. This first record was accepted only after I produced a colour photograph of the cock at the nest.

To the rearer of game chicks the sparrowhawk is the devil in feathers, the sparrowhen being the she-devil. Being strikingly bigger and stronger than her mate she can take bigger prey, including pheasant poults. As a result sparrowhawks are killed, or their nests shot up, protection notwithstanding. But not in

State Forests. There, like other birds of prey, it is safe, for the Forestry Commission's instructions to its officers at all levels are printed bold and clear.

Go into the wood where a sparrowhen has small chicks in the nest, and she will be off to challenge the intrusion by vocal display: *Kek-kek-kek-kek-kek-kek!* The chattering cry incises the quiet of the wood – sharp as a razor, challenging. The she-devil is as sharp as her voice, razor-faced, sharp-shinned slim-taloned, with yellow jewel eyes. She flickers round the slim larch tops with the agility of a swallow, negotiating with almost bat-like precision the fuzzy strata of tufted twigs. Then she will alight on a high twig and machine-gun you from there.

Two-thirds, or thereabouts, of the sparrowhawk's prey, are birds, with the other third made up of small mammals and insects. I have seen a small weasel brought to a nest. The cock, who does all the hunting in the early days, plucks the prey before passing it to his mate, who carries it to the nest in her feet. If she happens to be absent at the time he will drop it in the nest himself. The sparrowhawk's plucking place, or plucking stool as we used to call it, may be an old stump or a rock or mound, but is just as likely to be an old nest. The nest of the year is used as a plucking place for some time after the young have flown.

The sparrowhawk is a master of the surprise attack, the strike from cover, following a hedge line for a distance then corkscrewing over to swoop into a group of feeding starlings or fieldfares, or lifting from one side of a hedge to the other to snatch some unsuspecting bird by surprise. This style of hunting is an everyday occurrence, but one doesn't see it every day, not even when one is working with a pair at the nest. I see it regularly in the lane from the road to my house where the hawk and I have had some near misses.

A hawk pursuing a single bird can get itself into all sorts of trouble, ending up in strange places — cut up, stunned or even dead. I watched one following a starling into a byre. The starling knew where the bit of glass was missing from the window, and flew through. The hawk hit the pane and fell, stunned and cut. My wife and I revived it and dressed its sore face. Another flew into the stable, through the open door, and hit the wall, breaking its neck. Another flew into a byre after a sparrow and ended up in the grip, stunned. When it came to, it flapped up the grip, clarting itself with dung and urine. In the morning it was like a plaster cast. My wife and I had to bathe it, scrape it, then gently coiffeur it until it looked like a hawk again.

Any place where small birds gather to roost will attract the local sparrowhawks. My bird-table is one such. Sometimes the hawk will come prospecting even before I have begun to load all the feeding gadgets with food. My table is placed hard against thick shrubbery making it easy for the birds to take cover from a frontal attack. During the winter of 1987/88 I had a few thousand starlings roosting round the house. The sparrowhawk pair contented themselves with living mainly off these. They stampeded roosting birds early in the morning by flying the flank of the shubbery, and the same in the evening when most of them had pitched. I reckoned that the hawks had two starlings a day each. Hardly a morning or evening went by without our hearing the cries of a starling in a sparrowhawk's claws. The only other bird I saw the sparrowhen knock down was a collared dove.

Scotland lost the osprey early this century, the last breeding record being in 1916. A pair bred at Lock Arkaig in 1908, after which only a single bird was seen until 1913. The Loch an Eilein pair bred in 1899, after which only a single bird was seen in 1901 and 1902. After that, nothing. But single birds were seen every once in a while on passage in spring and autumn. Now, by a remarkable turn of events, more people have seen an osprey, at the nest or fishing, than have seen any other bird of prey, except perhaps the kestrel.

This has been due, arguably almost entirely due, to the work of the Royal Society for the Protection of Birds, which has made the Loch Garten nest world-famous and set a pattern for other countries. At Loch Garten visitors can view the birds at the nest from a special hut, without fuss but under strict control. The ospreys using this site are the most closely studied in the world. By 1988 the number of visitors reached 1·3 million. During these years ospreys were colonising elsewhere, unpublicised. In 1987 the number of ospreys successfully reared was fifty-two; the number for 1988 was seventy-five. So the 100 mark isn't a long way off. The osprey is no longer a rare bird in Scotland. There is every likelihood that it will soon return to England again.

I saw my first osprey in 1929, in Spain, which had a lot of ospreys when Scotland had none: now Scotland has more ospreys than Spain, where they reckon that the Balearic birds are down to ten breeding pairs. Factors in this decline have been named as tourism, disturbance and marine pollution.

The osprey, as any photograph or painting shows, can hardly be mistaken for any other bird of prey in this country. No doubt about identity remains when you see a big hawk, dark above and white below, plunge into the water and catch a fish. Long claws, and spines on the underside of the toes, give the bird a good grip. Usually it flies with the fish held head forward. It will trail its feet in water to clean them, and maybe duck its head as well. When roosting it spends much of its time preening and oiling itself.

These days, the only enemy of the osprey is that quaint breed — the egg-collector.

Starlings return to their nests as soon as the weather improves in January or February. They chortle and squeak and imitate other birds. Some starlings imitate the trill of a Curlew making one think the Curlews have returned early.

# I TO THE HILLS

David Stephen with his grandson and a baby roe deer.

There is an old Gaelic rhyme that reads like a statement of fact, although it is near enough only up to a point, and mostly wrong thereafter:

> Thrice the age of a dog the age of a horse;
> Thrice the age of a horse the age of a man;
> Thrice the age of a man the age of a stag;
> Thrice the age of a stag the age of an eagle;
> Thrice the age of an eagle the age of an oak tree.

The milk-white hind of Loch Treig didn't live for 160 years, despite the fact that she was never fired at. Nor did the great stag of Badenoch — the Damh Mor of the legends — have a lifespan of two centuries, however great and magical he was. The life of a stag can be measured by the rise and fall of his antlers, as a salmon's is by its scales, or a tree by its rings of growth. At twelve the Highland stag is in his prime; thereafter he begins to decline in prowess and growth of antler, and is old

*opposite.* Red deer stag in velvet.

before boys have attained manhood. A score of years he may have, but he will not be seeing the hills as he did.

In the wild there is no place for the aged or the ailing. In the days of the Great Forest, when Scotland was a kingdom, the wolf preyed on the herds, as wolves today prey on the caribou of the tundra, keeping them fleet and strong, so that death by time or misery was a rare tragedy. Today, man kills the red deer selectively. The stalker, treading the remote places with glass and rifle, grasses with a bullet the ailing and the old, the injured and the unsightly. And in due season the herds are culled, about 35,000 of them, for trophy heads and food as well as in the interest of the deer themselves. There was a time when you could hardly give venison away; now the red deer is the animal with the golden hoof, and deer farms are springing up all over the country.

At birth the deer calf, hind or stag, is an awkward sprawling bundle of legs, unable to stand or lift its sleek head; wet and squirming, snorting and choking with the gleet in its mouth. In that first hour the hind stands over it, licking it with her gentle, soothing tongue until its spotted coat is curled and dry. In the early days the calf lies still where the hind puts it down — chin to ground, unmoving except for a tremor over ribs after a deeper breath, or the opening and closing of nostrils gathering the scents of its new world. It is a beautiful creature, the calf of the red deer, with eyes glowing without highlight under sweeping lashes.

But the days of weakness are brief, and when the calf is running with its mother it has little to fear except from mountain fox or eagle, and the hind doesn't make access to her calf easy for either. Even when she is some distance from her calf it will usually be with others under the watchful eyes of an old aunt. The red deer society is a matriarchy, with Landseer's monarch of the glen merely a constitutional monarch for the short period of the rut. But the hinds make the decisions, and if the leading hind decides to go somewhere the others follow, and the stag has to tag on or lose them.

As a yearling the calf will still follow its mother, even when she has her new offspring at foot. A stag calf reaching the age of three may still be running with the hinds — immature, a knobber, unstirred yet by the fire of September/October. Among deer the knobber is still a growing boy, and in the forest, where there is an ethic in shooting, he is not considered shootable. At

Two young male roe deer
in a young forest
plantation.

three he becomes staggie, and any time after that he will join a
stag group. At six he will be a stag, but his antlers will still be
light, a promise of things to come. He may grow into a good-
headed stag, or a switch (with no points above his brow tines),
or a Royal, with brow, bay, and tray tines and three-point top,
or he may grow nothing at all and become a hummel, a heavy,
powerful beast unloved by men.

When first he takes the rut it will be as a flanker, a
skirmisher, a raider of harems, a cutter-out of straying hinds.
Only when the days of roaring are far gone will he try to play
the master and rut the last hinds of the season. But his day
comes at last, when the fire kindles in him, and he is in his
prime. Clear, vitreous blue are the day skies, with the wind snell
from the north; at night the peaks are etched, indigo and
glowing purple, against fiery sunsets, when the peat pools,
brimmed, are pools of fire on the shadowy levels. And when the
moon rises, frosty-brilliant, the hills are a spectral wonderland,
mirrored dark on silvery waters.

These are the days of roaring. Now he bursts into hind
ground — maned and swollen of neck — to rip heather with his
antlers, to wallow in peat hags and rise up like some primeval
monster dripping sludge and water. He roars and grunts and
challenges. He collects a harem. Rivals challenge him and he

Red deer stag in July just after cleaning velvet from the antlers. It is a 10 pointer and has probably been fed in winter. Now it luxuriates in rich pasture although as the rut (mating season) approaches it will probably head for the hills.

Red deer stag in velvet.

meets them in combat, which is mostly an antler-to-antler shoving match with victory going to the stronger shover. There is much sound and fury but little blood spilt. Sometimes fights are long drawn out; sometimes a stag is injured; sometimes the injury is serious. Once in a while the combatants get their antlers locked and have to be sawn apart. And that can be dangerous work for one man. It is safer with two men, freeing the stags simultaneously.

The rutting stag isn't dangerous; he will run rather than loiter. In any case the hinds go at once when disturbed and he has to follow if he wants to keep them. But a stag can be attracted closer if one's presence isn't suspected. A stalker did it for me when I was photographing. We stood up when I had finished but the stag kept coming, and we had to wave and shout at him before he turned away and fled in pursuit of his hinds. A wounded stag on his feet will always try to go; a wounded stag down can be dangerous when he throws up his head or lashes out with a hoof.

The really dangerous stag is the tame one, when he is in hard antler. My own big stag Cruachan scooped me up once and threw me on to my back, and he would assuredly have crippled or killed me if colleagues hadn't come to my rescue. As it was he split my upper lip like a hare's, broke some of my front teeth, leaving me like Dracula dripping blood, and grazed my femoral artery.

A tame roebuck is just as dangerous despite his small size, and his antlers are real stabbing weapons. The behaviour usually begins when the beast is in hard antler, but I knew a yearling, with his first antler buds, who pushed a postman into a burn. I have had several more or less bloody encounters with tame roebucks over the years, but have reared only two by my own decision.

The others were brought to me by people who had found them 'abandoned'. If the fawn was picked up that day I told them to take it back and put it down where they had found it. If it had been two days more in their possession I told them to take it to the SSPCA and ask for their advice. I used to take the fawn if it was a doe, and tell them to rear it themselves if it was a buck. In fact, I used to take either, after giving the 'rescuer' and earful of sound if angry advice. Roe fawns, like leverets, owlets or anything else, become 'lost' only after they have been found 'abandoned'. The golden rule is to leave all young wildlife alone.

The notion that the roe doe abandons her fawns arises from her habit of leaving them unattended between sucklings, during

Red deer in velvet.

the early days of their lives. They are safer lying hidden, mute and unmoving, until they are strong on their legs and able to run with her. After being suckled the fawns are put down again, but they don't huddle together: they lie apart, often some distance apart. When the doe returns again she calls to them (*whee-yoo*) They answer with a *peeping* sound and come running to her. I once watched a doe with twins giving suck to a third fawn whose mother had been shot, and which had passed my back door crying for her.

Early on at Palacerigg Country Park I reared two fawns, a buck and a doe, from different parts of the country. Bounce, the buck, grew into a handsome beast, and put up six points in his first year. The doe, Sanshach, produced twins for a few years, then surprised us all by giving birth to triplets, all of which she reared. That buck remained friendly with my wife and me, and my big German Sheperd bitch, throughout his life and made no truculent demonstrations towards anyone in our company, but I didn't allow him visitors in July or August, just in case. The rutting season for roe is from mid-July to mid-August.

The roebuck is the small-antlered deer. His antlers rarely exceed nine inches in length, although I have found cast ones up to ten and a half inches. Bounce grew strong, heavily pearled antlers of that length. Very infrequently a doe will grow short antlers which remain permanently in velvet. Once his pedicles have sprouted the buck fawn may first of all grow a small, horny 'button' on each tip, which will fall off in January or February, or he may go right ahead and grow antlers. His first growths may be single spikes, or forks, or more rarely his full six points.

In roe deer there is delayed implantation until the end of the year, but this will vary according to whether the doe was mated earlier or later. Most births take place in late May or early June, but my wife and I once watched a doe birthing on the fourth of July. During the rut the buck runs the doe on rings, which are well trodden and easy to recognise. The rings are always round some natural feature like a tree, or bush, or boulder or a knoll. Does and fawns run these rings at other times, but nobody yet knows why they do it.

Roe deer are territorial. The buck marks out the boundary of his territory by setting scent, by thrashing vegetation, leaving scent on it from the gland between his antlers, and by scraping the ground with a fore hoof. He defends his territory aggressively, barking much and driving off any intruding buck. A challenger, or an intruder, isn't usually pursued beyond the boundary of the territory. Fierce fighting often takes place, and many a loser leaves with a bloody nose or a flank wound. Fatalities are rare, and I have only once seen one buck kill another, by accident. The doe will drive off intruders from her ground before and after birthing.

Before high seats became the vogue for watching roe deer I found out their usefulness by chance. One morning, after spending the night on a tree platform photographing owls, I watched a roe doe and her two followers browsing in the wood. On another morning a buck came and rubbed himself against my tree. I left some of these platforms up for some years afterwards so, in a way, I can fairly say that I invented high seats. From such platforms, in later years, I was able to watch roe deer running the rings.

The fallow deer is lighter in weight and shorter in stature than the red, taller and heavier than the roe, but a good Sasunnach buck will weigh as much as many a poor Highland stag. The fallow buck is the only British deer with palmate antlers, and the species the only one spotted with white when

Roe deer just having had twins, an unusual occurence.

adult. But fallow come in several colour phases, from near-white to near-black. The standard is fawn with white spots in summer; in winter the coat is greyer and the spots faint.

It is now generally accepted that the fallow deer of today are the descendants of animals imported from southern Europe. We tend to think of the fallow as a park deer, an ornamental species, and certainly most parks have them. In the neighbourhood of parks there are usually animals that have escaped. So fallow deer are liable to turn up in unexpected places, as on the old Glasgow to Edinburgh road near Castlecary. But truly wild fallow deer we have too.

In some places, like Perthshire, Ross-shire and Argyll, there have been wild fallow for a very long time, and probably for two centuries in the area of Dunkeld and Blair Atholl. They are present in south-west and north-east Scotland, and on Mull, Islay and Scarba. Anywhere between Clyde and Forth one might meet up with fallow deer. In 1951, when I was investigating deer poaching, the Marques de Torrehermosa told me that his herd of fallow had been wiped out by a gang using Sten-guns.

Fallow deer like plenty of cover, and truly wild ones are shy and secretive. Although they are mainly active from dusk to dawn, they will feed and lie out in quiet places by day. Old bucks are said to be the most strictly nocturnal. Female fallow bleat, fawns bleat, and bucks during the rut belch loudly. This is generally referred to as groaning. It has never sounded like groaning to me.

The sika deer is an introduced species, its Latin name *Cervus nippon* indicating its place of origin. Weighing less than the best fallow, it can match the not-so-good. The summer coat is chestnut, with spots that are more or less white; in winter the coat is grey-brown with few or no spots. Calves are spotted at birth. The sika stag's velvet is black, with red patches on the beams. Sika are well established in Argyll, Caithness and Sutherland, Inverness-shire, Fife, Peeblesshire and Ross, and in the Highlands there has been some hybridising with red deer. Unlike the red stag the sika doesn't roar; he whistles and creaks. Again unlike the red stag the sika is tolerant of lesser beasts on his ground. Like the red deer, sika live in hind groups and stag groups. These groups feed mainly at night, especially when clear of cover. The sika stag is polygamous. The rut begins at the end of September, and calves are born late in the following May or in June.

The reindeer was brought to Scotland by Mikel Utsi, with whom I spent a few days after the first arrivals. He was of the opinion that there were no reindeer in Scotland before the viking invasions, so the beasts that became extinct in the twelfth century were the descendants of imported stock. This view is being held increasingly today, because of lack of evidence of the reindeer's survival into the post-glacial period. The Scottish herd is located in the Cairngorms. In the reindeer both sexes are

Reindeer stag on the slopes of the Cairngorms in winter.

antlered. The calves at birth are brown without spots. The Cairngorm herd is managed, and the beasts come to a call. Mikel Utsi used to give children in Aviemore a great thrill by arriving with reindeer and sleigh, dressed as Santa Claus. Millions of children have met Santa Claus; not many can say they have patted real reindeer.

There is no such thing as a native wild goat in Scotland. Today's feral herds are the descendants of domestic goats gone wild and have probably changed little in appearance over the

Mikel Utsi feeding reindeer. This Finnish reindeer manager introduced a small herd to the Cairngorms in the Scottish Highlands.

centuries. They come in a variety of colours and markings: black-and-white, brown-and-white, black, brown or near-white. They are small, hairy and hardy, and in most places lead a spartan life. Kids are born early in the year, January to February, and into March, and mortality can be high in severe winters. This makes for rigorous selection. Apart from culling by weather, there is some predation on kids by foxes and eagles.

Some herds have been killed out; some are culled. The Forestry Commission was compelled to make a drastic reduction in the goat population of the Galloway hills. The main herds are in the south and west of scotland, including some of the islands. Generally speaking the herds can be viewed from quite close quarters — I have had nannies with kids walk past me at a distance of a few yards — but where they are under pressure they can be as wild as deer.

I introduced a small herd to Palacerigg Country Park, the founder member being a billy kid from Ben Lomond. His mother had been killed and he was hand-reared until weaning when he was passed on to me. I called him William. When he was fully grown I brought in two nannies, and in due course there was a group of five — William, two nannies and two kids.

William liked to rear tall then butt down to slap my palms with his horns. One morning, when a fox entered the field, and began to stalk the nannies and kids, William reared tall, butted down and charged. He chased the fox to the fence and butted

William's wife and kid at Palacerigg. Today's feral herds are the descendents of domestic goats gone wild.

up at him as he went over the top. It would be nice to think that William was defending his family. More likely he just didn't like the fox.

The other beast of ancient lineage that I introduced was the Soay sheep, which is a primitive domestic type. It was found only on Soay until it was introduced to Hirta. Now there are many small flocks on the mainland, some of the dark brown type, some of the fawn type, and some mixed. I chose the dark variety.

The phrase 'never in a thousand years' has meaning when applied to the St Kildans and the Soay sheep. After a thousand years or more of living together the sheep refused to be dogged and the islanders never could breed a dog that could work them. At sight of a dog the sheep broke away and ran for their refuges on the cliffs. Doctor Morton Boyd has told how a first-class mainland sheepdog, working for a scientific party on Hirta, completely failed to herd them. My small Palacerigg flock accepted as friendly my German Shepherd bitch and the lambs put up with her sniffing them.

The St Kildans trained their dogs to run individual sheep into a deadend and hold them there. Most Soay sheep today will run at the sight of a human being. Those living near the old village have become accustomed to the new residents and tolerate them down to about ten yards. The Soay sheep is a unique breed, which has remained pure despite the fact that the St Kildans introduced other breeds and crosses over the centuries.

# LA CHASSE

Capercaillie hen on nest in pine forest. They will stay put until almost trodden on and most nests are found by accident by flushing the sitting bird.

*Tick! Tick! Tick! Tick!*

It is a bird ticking, not a watch. I strain my eyes in the gloom of the pinewood — it is half past two in the morning, with a fine smir of rain like gossamer — and presently I see the dark bulk of him, parading round his pitch below a knoll: the biggest grouse in the world, beaked like an eagle, booted like a grouse, with his tail fanned, his head erect and his beard on end — the *cabhar coille* (old bird of the woods) of the Gael, a cock capercaillie.

He strutted and slow-walked, clicking his beak and *ticking* his halfpenny watch. Every now and again he would leap into the air like a blackcock, and sometimes he would give a cat *squech* or pop a cork. There were other cocks parading round the knoll, posturing and challenging, but none came to blows that morning. Indeed, the display of the cock capercaillie is largely formalised, like the opening of Parliament. Fierce fighting, however, does sometimes occur.

The capercaillie became extinct in Scotland around 1760, but was successfully reintroduced in 1837, and in later years. Since then it has colonised north, west and east, and as far south as Lanark and Dumbarton. Colonising is led by the hens which may not be followed by the cocks until a year or so later.

*opposite.* Cock Capercaillie displaying his fine feathers.

105

During that period the hens will breed with blackcocks, and at one time such hybridisation was not uncommon. There is a record of nesting at Auchengray Estate, a few miles from the town of Airdrie, in 1916. When I asked the retired keeper about it in 1930 he expressed the view that the hen capercaillie had been mated with a blackcock

Sometimes a displaying cock capercaillie makes the news, as when he attacks people, Land-Rovers, sheep, keepers and postmen: a slightly mixed-up bird who doesn't know what he's doing. One that attacked me in Glen Esk was killed doing battle with a van.

The hen usually scrapes her nest near the base of a forest tree, or stump, but some will do so among brushwood or in heather outside the main tree cover. Most birds will sit until almost trodden upon, and I should think that most nests are found by accident, by flushing the sitting bird. A man, covering the same ground every day will tell you that he sees a dozen hen capercaillies for every nest he finds, and that his dog will put a hen off only once in every two years or so. The broody hen is a ponderous, slow-moving bird, who can sit out the daylight with hardly a movement, except when she stirs to turn her eggs. When her chicks hatch, she leads them away into the forest after their down has dried.

The capercaillie in a tree is a far more alert bird than one on the ground. When disturbed at roost the cock goes rattling through the trees like an express train, showing remarkable agility for such a large bird.

Like the cock capercaillie, the blackcock has a specialised display, the area where it takes place being known as a lek. Lekking blackcock are not difficult to watch, and I usually put up a hide on the edge of the display ground to let the birds become used to it before they begin their performance. At one lek where I spent a lot of time the birds usually arrived at 2.15 a.m. (British Summer time, early May). They came in with a great flurry of wings, barely visible in the dark, sorted themselves out on their stances, and the singing began:

*Roo-roo-ruckoo-coo-roo-cucu-roo.*

As a spectacle the display of blackcocks is fascinating; understanding the significance of the movements, and the strutting, and the comings and goings was puzzling. Marking of birds in recent years has shed much light on the subject, but there is no room here to deal with the complexities.

At the lek the birds sing on their stances, with heads down and necks inflated like pouter pigeons, and they leap into the air, crowing. The crow sounds to me like *co-whae*. It may not sound like that to anyone else. Every now and again a bird will strut from his stance, with lyre tail spread, and advance to meet a neighbour at the boundary. They advance and retreat, or parade sideways, or leap to the clash, making a few feathers fly. During a clash the birds utter a variety of sounds, throaty, stuttering and cat-like. Sometimes three birds will rush towards each other, but usually one stands aside, leaving the confrontation to the other two.

The female black grouse is called a greyhen. The greyhens visit the lek in the morning, and are mated there. Their arrival causes a great burst of activity among the blackcocks. They leap and crow and make short flights all over the place. This excitement lasts for only a short time, the birds rising and touching down, again and again, like a lot of pinioned birds trying to fly.

Sometimes one comes across a solitary blackcock *coo-rooing* on the ground far from the lek, and sometimes one finds one doing so in a tree. I used to have a quiet laugh at such birds until I saw a greyhen associating with one on the ground. Blackcocks don't need a lek when they mate with hen capercaillies. So why not with a greyhen off the lek? After mating the greyhen is on her own. She scrapes her own nest, she sits on the eggs and she rears the chicks, all without benefit of a mate.

The red grouse was, and is, an important bird in many ways. For a hundred years and more it has been responsible for a particular form of land use, involving the rotational burning of large areas of heather moor. It has been a source of large rents to moor owners and of revenue to Local Authorities. It attracts great numbers of people from overseas who are prepared to pay large sums of money for the privilege of shooting it. It gives seasonal employment to many people, and full-time employment to others, such as gamekeepers.

It is a managed species, with man manipulating the habitat, which is heather moor. The red grouse lives in the heather, it feeds on heather, and it nests in the heather. The area of heather, and its nutritional value, determine how many grouse a moor can carry. Heather rich in nitrogen and phosphorus has been shown to be a major factor in population density. Without heather there are no grouse. Birds driven to the fringe of a

Blackcocks displaying in spring on the Lek (display grounds). The display is all show and the birds rarely come into serious conflict.

moor, or off it altogether, do not survive for long. They die by predation, starvation or disease.

The cock grouse is territorial, defending and holding his ground by aggressive display. The size of his territory is related to the quality of the habitat — which is the heather. Once the territorial birds have taken their places, all the birds without territories are banished to the fringe of the moor or to patches of poorer ground within it and not occupied or defended. In winter the death rate among these displaced birds is high. Any vacancy on a territory, by the death or disappearance of one of the pair, will be quickly filled from without.

In this way the red grouse regulates its population density, holding smaller territories on the best ground and larger ones on the poorer. On a moor where all territories are occupied it follows that every chick reared is surplus to its carrying capacity. Good management of heather determines carrying capacity. And the bird's social status is important to its survival. Being a territory-holder gives the bird an edge, so to speak. The

vulnerable ones are those off the moor or on poorer ground.

Since the war the red grouse has been declining for one reason or another, and sportsmen generally still like to hanker after predators as the cause, although it has been shown again and again that the number of grouse determines the number of predators and not the other way round. In any case all our birds of prey are now protected by law. But there are still plenty of lairds and keepers around whose fingers are itching for the trigger or straining to uncork the poison bottle, as the Game Conservancy, in a spirit of sweet reasonableness has admitted. These people are still pushing for an open season on predators of grouse, which, despite eagles and falcons and harriers, peaked in 1987 as predicted by Watson and Moss of the Institute of Terrestrial Ecology in 1985.

When I find three or four brace of grouse in an eyrie, I know at once, as anyone should, that there must be a lot of grouse around, and that the prospects for August are good.

The ptarmigan is the grouse of the high tops and the

Red Grouse on heather moorland. Grouse stake out a territory which they defend, and breed in.

snowline; on the high tops most of the time and on the snowline in winter. It takes a prolonged blizzard or lasting snow and ice to drive it down to the glens. But the birds are quick to move up, or ahead of, the retreating snowline. In their plumage of winter white they are almost impossible to see above it; below it they become strikingly obvious against the dark ground. Yet there many of them feed, while others, higher up, are scraping into loose snow for a bite.

Generally speaking, ptarmigan are found from 4000 ft down to 2500, but here and there, as on Islay, they come even lower, to the 2000-ft contour. Nests are usually between 2200 and

Heather is everything to grouse — food, shelter, escape, cover.

4000 ft. The hen makes a simple scrape among stones or rocks, or plants like heather or crowberry, or in the shelter of a big boulder. On the nest she so matches her surroundings that she is difficult to see, and some birds will sit tight until almost trodden upon, or even after being almost trodden upon. A flushed bird doesn't usually fly very far from the nest. Some birds have a high tolerance of a human presence; others are more shy. Any time I want to return to a nest I make a little cairn of stones nearby to mark it, unless it is beside some obvious natural feature like a big rock.

Up on the high alpine plateaux, the cocks are fond of sitting on little prominences while their hens are sitting on eggs, and you are as likely to hear ptarmigan-speak before you see birds as the other way round. The usual sound is a croak, but the birds can string notes together into a crackling chain-song, likened by the late Seton Gordon as 'not unlike the winding of a clock or the ticking of a fishing reel'. This is a good description, but the sound is more rattling to my ears.

The ptarmigan's flight is remarkable. It is the typical game-bird's flight — rapid wing beats with periods of tilt and glide. But it has other characteristics. The bird swings about in flight, like a motor-cyclist on a racing S-bend, and it heels over in much the same way. But the most notable thing to me is its ability to rise vertically when a cliff or steep slope gets in its way.

There is no hard evidence that the Romans brought the pheasant to England, where it certainly arrived before the Norman Conquest. It didn't arrive in Scotland until the sixteenth century. Unlike the alien rat, rabbit and grey squirrel it requires no keeping down; the problem has always been to keep its numbers up.

Two world wars have made it clear that the pheasant is able to shift for itself, despite no keepering and the presence of a predator force. But, unaided, it will always be thin on the ground. The big battalions have to be produced by man, and made up by him each season. A big show of pheasants depends on artifical rearing and a lot of expenditure by those who shoot.

Many people think it is shameful to rear half-tame birds then turn them loose to be shot later in the year for fun. I am no partisan of game shooting, but I think this is more defensible than sticking the birds in battery cages or broiler-houses, which is the way most people get their eggs and chickens nowadays. Large numbers of chicks are bought from Game Farms, but some of the big estates produce their own. The keeper catches up cocks and hens, confines them until the hens have laid all their eggs, then turns them loose again.

The first pheasant to be naturalised was the black-neck (*Ph. c. colchicus*). In the eighteenth century came the ring-neck (*Ph. c. torquatus*). The ring-neck passed on its ring, as surely as a Hereford bull passes on his white face, and practically every wild cock pheasant has a ring or a white mark of some sort on its neck. You can turn down black-necks any time you like but their characteristics are soon drowned in a river of mixed blood. Oddly enough, though, there are still pure black-neck cocks in Palacerigg Country Park, where I introduced them in 1974.

Shooting men often say (perhaps in unwitting defence of hand rearing) that a pheasant is wild from the day it flies. This is nonsense. You can keep a pheasant tame after it flies if it was tame before, hand tame that is. I had one ring-neck cock who was as tame after a year's flying as he was before his first true flight.

Hen Pheasant.

*opposite*. Tame roe deer with one of the Stephens' Rhode Island Red hens.

Walking up partridges is common practice when you're shooting them; walking *with* them is something else. I've done it with only one brood of partridges in my life, and that came about because a fox killed a hen partridge when she was off her nest one afternoon. The 16 eggs in the nest were cold as marble when I lifted them. I gave them to my broody Old English Game bantam, Spick, who was sitting on a nest in my back porch.

My big Labrador bitch, Tarf, spent hours sitting beside Spick, periodically taking a sniff at her, when she would fluff up her feathers and croon. The two were old pals and Spick was as often in the house as out of it. Spick hatched all 16 eggs. For two days I kept her in the porch where the chicks spent much of their time running over the dog like fleas. So they took her for granted when I let Spick loose with a following of 16 cheepers and a big Labrador retriever. When Spick settled to brood the dog bellied down beside her.

Tarf the Labrador was taken aback when the partridge cheepers made their early precocious flights. Within the week she was even more surprised when they went whirring over the hedge, leaving Spick running this way and that and trying to take off after them. They soon came back, not flying, but via the bottom of the hedge. When they were strong on the wing, and jugging outside, I shut Spick up at night. Early in the morning I let her out and the partridges soon found her.

When I went walking with Tarf the birds would sometimes walk ahead, talking and nodding their heads as partridges do; at other times they would come hurtling low over her head, forcing her to duck; sometimes they came down in a cascade about her ears. When the cornfield at my back door was cleared they spent a lot of time on the stubble. So did another covey of partridges. I always knew my birds because of their behaviour towards the dog, and I began to wonder if they would come to me if I walked alone. I walked alone and they didn't come. I walked with the dog and they did.

Before that cornfield was cut it had a strange visitor. When I came home after a few days' absence my farmer neighbour told me there was a strange bird calling there in the evenings. It had arrived on the day I left. I went out with him and heard the *Wet-my-lips* call of a quail, a voice I hadn't heard for 21 years. My neighbour had never seen one, and wanted to know how to go about it. I told him that it would be as difficult to find as the elusive corncrake.

Next day I saw it twice, in my stackyard, where Spick was scratching around with her partridges. Several times the bird approached them and it seemed to me to be trying to cut some of them out. I chased it back into the cornfield. That night there was steady rain and I had a call from my neighbour, who was soaking wet. 'I've something for you,' said he triumphantly. He had. He had stalked and caught the quail! He had wanted to see one, and he did. I photographed it before he took it back to the edge of the cornfield.

Spick, The Rhode Island Red hen, poaching from the garden birds!

# HEDGEHOGS, MOLES AND SHREWS

A shrew. These tiny animals need greater quantities of food: up to and even more than their body weight each 24 hours.

Although it belongs to the order of insectivores the hedgehog takes a lot of other prey besides, and tame or captive ones will eat almost anything. On two occasions, years apart, I found a hedgehog eating a frog, the first crouched with a dead one in its teeth, the second eating one from a hindleg forward while it was still alive. I have seen one at my quarry pond in spring, pottering about on the edge of the shallows, when the frogs were spawning. On another occasion I came on one feeding on a half-eaten frog, which it may or may not have killed. One of my tame hedgehogs tried to catch that year's tiny frogs in my back porch but they were much too nimble for him. I would imagine that hedgehogs lose more frogs than they catch.

What a tame hedgehog will eat isn't necessarily a reliable guide to what they do in the wild, but if one of mine will eat slow-worm is it a fair assumption that a wild one would eat such a prey if it could catch it. The same applies to nestling mice and voles and, for that matter, small naked rabbit kits. I once found a rabbit stop open by day, and thought that the young had left, until I reached in with a hand and felt the spines of a hedgehog. Tame hedgehogs eat dog- and cat-food; so will wild ones if they are baited to it and you keep putting it out for them.

Hedgehogs can, and do, kill adders but nobody knows how

*opposite*. Hedgehog exploring a cairn.

117

often they do it. Maxwell Knight recorded that they prefer adders to grass snakes. One of my hedgehogs, confronted by an adder, coiled into a ball and refused to budge; another crouched, with brow quills forward, and let the snake strike at his armour, then crawl over him and away. Many people still believe that the hedgehog is immune to adder venom. It isn't. One of mine, struck on the face, died within 36 hours.

The old folk-tale about hedgehogs sucking cows dies hard and a hedgehog couldn't suck a cow if it tried, and no self-respecting cow would permit it in any case. In 1963 I wrote about this in my book *Watching Wildlife*: 'What the prowling hedgehog does find are trickles of milk pressed from an overstocked udder when the cow is lying down. The globules are held for some time by the moist, pressed grass. Many years ago I noted, and wrote about, this.' When I lived at Luggiebank I spent a lot of summer nights among my neighbour's dairy cows, and sometimes there would be a hedgehog in their field. Some cows rose, snorting steamy breath, when the hedgehog came too close; others were unconcerned. One cow got to her feet when a hedgehog came too close to her udder. My torch lit up milk globules on the grass.

The hedgehog builds two kinds of nest—a big one for birthing in, and a big one for hibernation. Both are built of grasses, bracken, leaves and moss, all carried in the mouth. The hibernation nest can be in a bramble thicket, under the exposed roots of a tree, in a bank hollow, in an old rabbit burrow or, in one case, under the floor of one of my hen-houses. At Firknowe I found one in a cavity of the old drystane dyke. The beasts lay on fat in autumn to fuel the long slow-burn of hibernation, during which heartbeats and respiration slow down and body temperature falls. My own hedgehogs went into hibernation in November and remained there, more or less, until the end of February or early March. The three in my cold greenhouse at Firknowe did a bit of swapping over, one emptying its box and rebuilding its nest in a corner behind some flower pots.

Wild hedgehogs are not difficult to follow when hunting if one takes care not to tread on their heels; tame ones are easier still, and I have crawled all over a field beside one. I don't see so many of them nowanights, because they are not as common as they once were. Their main predators are man and his motor vehicles. Caught in the headlamps the beast assumes its usual defence posture of rolling itself into a ball, but its hedgehog

Mole with earthworm, the preferred food. Moles immobilise worms with a bite at the fore end so that the worms don't go off in storage.

defences are no defence against the wheels of a truck or motor car.

The mole or moudie, the Jacobites' little gentleman in black velvet, spends almost its entire life in underground tunnels and isn't often seen on the surface alive. Of course it has to surface to drink, from a ditch or puddle, and tunnels a branch line to the supply, but not many people have seen one drinking. For that matter, not a great many people have seen a mole. A moleheap a few feet away from my small garden pool often attracted my terrier who would stand over it with her head cocked, lending an ear to movement in or below it. It was assumed that the mole was drinking from the pool. It certainly came to the pool. I know, because I found it drowned there.

In captivity the mole sleeps between bouts of eating. Modern research has shown that, in the wild, it works roughly on a four-hour-on and four-hour-off basis. It has to eat more than half its body weight every day to stay alive, so it hunts by day and night, throughout the year. When I was at Luggiebank we found a store of earthworms about a foot below ground in the corner of a garden frame. All the worms had been crippled by a bite at the fore end. The mole had a tunnel under the frame, connecting with the nearest moleheap. We happed its cache up again.

My old Labrador Tarf was an expert at scooping moles from moleheaps and was my main supplier of live ones, which she harmed not at all. In recent years I have used plastic mole-traps in my garden. The mole walks in, the flap closes behind it, and it is caught. I put some cat-food in the traps so that the mole wouldn't go hungry between my visits. I once caught a small bitch weasel in one of these traps. My Siamese cat killed the occasional mole in late summer, but didn't eat any of them. Spider, my tawny owl, had no such reservations and ate any small mole he had caught or was given.

Shrews belong to the same order as the hedgehog and the mole, although they all look like mice and, indeed, were referred to as such by my elders and the boys of my generation. Screw-moose was the name we gave to common and pygmy shrews. For the water shrew, the biggest of the three, we had no local name that I can remember, and it was the water shrew to me when I first met one in my early days at Secondary School.

Pools and burns of clear, unpolluted water are the preferred habitats for the water shrew, although it will sometimes be found living and breeding in woodland far from either, making do with ditches. This is usually a black and white shrew, black above and white below, with a clean line of demarcation along the flanks, but in some animals the white becomes greyish, and the black has a flush of dark brown. On the underside of the tail there is an obvious keel, or fin, of swimming hairs. Overall length is 4–6 in., of which 1·8–2·4 in. is tail.

All the shrews are strong for their size, and this is the strongest, being able to kill small fishes and frogs, and very big snails. It is often accused of eating fish ova, but its predation can be nothing but insignificant. In turn it is preyed upon by the pike, and Jim Lockie found one in a pike's stomach. Tawny owls kill them and certainly eat them. So do barn owls.

The water shrew is an accomplished diver and swimmer. It swims high in the water, with lateral movements of head and body, and paddling with alternate strokes of its feet. Air bubbles, trapped in its fur, often give it a silvery appearance when it is swimming under water. Like the dipper it can hunt the bottom, wiggling along like a newt, while it turns over pebbles with its forefeet searching for caddis grubs and other prey. I kept two young ones for a few days when I was at Firknowe, for the pleasure of seeing them diving in a big fish tank to pick natural prey from the bottom.

Not long after I moved into my farmhouse at Luggiebank I had a call from the miller saying that his terrier had uncovered a nest of wee beasts in the bank beside the narrow channel feeding the dam above the millwheel. There were five small young in the nest, which were not moved by the mother although the site had been almost uncovered by the miller's terrier. The nest was a ball of leaves and grasses, with several tunnels leading up to the top of the bank and one leading down to the water. The water shrew had a number of feeding places along her hundred yards or so of channel, and at them we found broken caddis cases, pieces of shell and other prey remains. Several times we watched her swimming the bottom, kicking up a cloud of ooze sometimes when she pounced. She could sink from surface to bottom like a dipper, and was no more difficult to watch once I had found her.

The pygmy shrew is the smallest of Scotland's three species; the common shrew comes between it and the water shrew. Except in size the two species are very much alike. The pygmy has a relatively longer tail, about two-thirds of its body length against about half in the common shrew. The pygmy is not the most numerous shrew species, but it is the most widely distributed, being found from sea-level to the tops of mountains, throughout the mainland, and on most of the islands.

Shrews live at high pressure and age quickly; most of them die before their second autumn. They are active and restless, full of nervous energy, ever on the foray, rummaging, scraping, burrowing and probing in the ground litter. They require great quantities of food, up to and even over their own body-weight in twenty-four hours. They die if deprived of food for more than a few hours, which is why so many are found dead or dying in vole traps. Foraging hungry after a rest the shrew will move a moleheap of litter in a few minutes of commanding gluttony. It rustles as it rummages, without thought of stealth, and squeaks as it rustles.

The pygmy shrew is an altogether more nimble and active animal than the common, as anyone who has kept them in captivity will know. It can out-run, out-climb and out-jump the common shrew, and in captivity keeps well out of the way of its domineering relative. Both have their likes and dislikes in the matter of food and, when given a choice, will make their choice. Earthworms are favoured; centipedes are low on the list. But a shrew will not go hungry rather than eat what it would reject when give a choice. I have kept both on a variety of food, including dog- and cat-food, hare, rabbit, cow and horse, with a little oatmeal (the last recommended by Crowcroft).

Mammalian predators kill shrews but rarely eat them. Dogs and cats that swallow them usually vomit them soon afterwards. Foxes, stoats and weasels can digest them, but appear to eat them only when the alternatives are shrew or hunger. Bird predators, on the other hand—tawny owls, barn owls, long-eared owls, short-eared owls and kestrels—eat them readily, and carry them to the nest when feeding chicks.

*opposite*. One of David Stephen's Tawny Owls.

# REPTILES AND AMPHIBIANS

The Adder, despite its reputation, is a shy snake.

I was looking at this adder, coiled beside a boulder on the hill path, and wondering what to do about him. For there was a group of a dozen or more walkers not much more than a quarter of a mile behind me who, for all I knew, might be of the breed that stands and stares instead of bashing and bruising, but I've seen so many adders battered to writhing death by twentieth-century St Georges that I decided to play safe and move him. So I hunkered down and told him so, knowing he couldn't hear but to get myself relaxed enough not to frighten him.

He was a somnolent in the heat of the sun, slack-wound not spring-coiled, relaxed in the wide-eyed, unblinking daydreaming of the snake, and I was sure he wasn't seeing me from behind his perspex guard because the head didn't turn to bring a look at me when I moved my hand gently in front of him. I came down closer to him and said: 'Hallo.' Then I lightly touched one curve of his body, and he gave a slight quiver, like a person

*opposite*. Frogs spawn in the same pond every year. Migration to the pond takes place at night usually when there is light rain.

125

stirring in sleep, then stared at me with his expressionless snake eyes, and I kept my face where it was to find out what he would do. He tightened his coils and raised his flat head a little, but made no move to retreat, and certainly none to strike. He was a placid soul. I ran a finger along his outer coil, stirring him into a gentle whirlpool of movement, and when his tail was clear I took it between thumb and forefinger and carried him to safety.

The adder is not an aggressive snake; it is, in fact a very nervous and timid one, wriggling wildly when first handled but easily soothed into quietude. True it is venomous, but fatalities are rare, around twelve in England this century and only one in Scotland in my lifetime. It is a fact that people recover from an adder bite who don't realise they have been bitten, putting the pin pricks down to something else, like gorse, dogrose, thorn or even bramble. Children are more at risk than adults, but all depends on the size of the snake, how much venom it has aboard, and how much it injects.

Adders hibernate, sometimes in groups, and it was when taking one from such a group that I was bitten in Spain. The males are territorial and fight chest to chest and chin to chin, sometimes crossing necks. This used to be called the dance of the adders. Mating takes place in May and the young are born in autumn. There is some association with the mother, for several days at least, and I have watched a big female basking in gentle sunshine with her small family. The young are venomous from the beginning, although their first prey is insects and other small fry. Adders do not swallow their young at threat of danger, or at any other time.

One of the unfortunate results of the general human hostility towards the adder is that the harmless little slow-worm is often killed in mistake for one. I have, in fact, received through the post a package marked: LIVE ADDER — OPEN WITH CARE, only to find inside a dead slow-worm. I wrote to the sender asking how he ever imagined the adder would arrive alive, and asked him not to do such a thing again.

The slow-worm is a legless lizard that looks like a small snake but isn't, and this snake-ness is its only resemblance to the adder. If you see something snaky and are in doubt, the snake is a slow-worm if it blinks its eyes. Adders can't blink. The adder is rough cut, rough hewn, and has a wasp-waisted neck. The slow-worm is smooth-skinned, highly sheened whatever its colour and has no obvious neck. Colour varies with age

Slow worms are neither worms nor snakes but legless lizards.

and sex, from various shades of brown to brassy, grey, chestnut, bronze or copper. Mating takes place in spring and the young are born usually in August or September. The female lays eggs, with fully developed young inside, which rupture soon afterwards and sometimes during extrusion.

There is a small white slug with a big name (*Agriolimax agrestis*) of which the slow-worm is especially fond, and in captivity it prefers this prey to anything else. On Ailsa Craig it feeds on the black slug (*Milax*) because the small white one with the big name isn't found there. Over thirty years ago, on one of my visits with my friend Jack Gibson, who was doing his annual gannet count, I picked up two slow-worms and brought them home. I gave them a choice of slugs. Once they had tasted the small white one with the big name they became saurian Oliver Twists.

The only other lizard in Scotland is the common one, just as often called viviparous because it gives birth to living young, which usually break out of their membranous sacs within an hour of extrusion. At birth they are under two inches in overall length, but are able to fend for themselves from the beginning, knowing what to hunt, how to hunt, and what to leave alone. They move about in short sprints, with sudden stops to listen and look about. They are not difficult to rear, and soon become tame, but I've done this only once because they are easy to watch in the wild as in a fish tank or whatever.

Common lizards are found over all mainland Scotland and on some of the islands, from low ground to the tops of mountains. There is a patch of rocky ground not far from my home where I've watched these lizards for more than fifty years. I sit down and wait for them to come to me; and they do. They run over my feet, climb on to my shoulder, sit in the palm of my hand, alert and bright-eyed, to bask in the sun. They don't like great heat, so Scotland suits them. Like the pipistrelle bat they will sometimes appear on sunny days in winter, even when there is snow on the ground. During the spring and summer they have favourite basking places, so you can have one beside you if you wish. They are curiously unconcerned about the human presence.

They become quickly concerned when the human presence chases them or makes a grab for them. Grab one by the tail and you'll be left with it while the lizard escapes. In due time the real bit of bony tail will be replaced by a gristly stump. The tail is designed for self-fracture. Even severe fright can induce it. Once in a while someone comes to me with a lizard, apologising for having broken its tail and asking if I can do anything about

Viviperous lizard, whose tail is designed for self-fracture and when it breaks off the tail, wriggles and distracts the predator.

it. I tell such people to put it back where they found it, and tell them about autonomy.

Boy and man I had a favourite quarry pond which I visited regularly throughout the year, and especially in spring, when I could see lizards, frogs, toads and newts without having to move more than a dozen steps in any direction. The man put away only some childish things, and still can't resist rearing tadpoles into frogs or making pets of toads. He also remembers the boy, so when he meets boys and girls at spawning ponds (or anywhere else for that matter) he is always happy to mix with them, talk with them, and be asked questions by them.

Do young frogs kill off old ones at spawning time? They don't, but the legend is slow to die. Usually the belief is held by old people, who grew up when it was widely held, and some of them pass it on. So I am still asked about it by young people of secondary-school age. A lot of frogs die, or are killed, at spawning time, but this has nothing to do with the young killing the old. A bull frog anchored on top of a big female — the amplexus position — is waiting for her to lay her eggs in the water, where he fertilises them. The eggs soon absorb water to become the familiar jellyfish masses polka-dotted with black.

Frogs spawn in the same ponds every year, and in the same place, and you can turn them around as much as you like without ever persuading them to change direction. How do they find their pond? Probably the same way as the salmon finds its home river — by smell. Migration to the pond takes place mostly at night, and mostly when there is light rain. At first there is a period of inactivity, which varies from year to year, then the croaking begins, signalling that spawning time has arrived. Toads using the same pond will spawn in another part of it, and usually in deeper water.

Does it ever rain frogs? This is another old belief which, although untrue, is based on the association between frogs and rain. Tiny frogs, which have spent ten weeks or so growing from tadpole stage, leave the water during rain, and during this migration to land may be seen swarming in fields, lanes and on country roads. If there is a high wind they may get tossed about like blown leaves, and some of them may arrive on your doorstep or in your porch, jumping about like fleas. In some years, at Luggiebank and Firknowe, we have such invasions in the porch, always on wet nights with a high wind.

Frogs go into hibernation in October or November, depending

on the mildness or severity of the weather. Adults will bury themselves in the mud at the bottom of ponds, which may or may not be the one in which they spawned, or in ditches, or in burn banks below water level, sometimes in the disused burrows of water voles. Young frogs will hibernate almost anywhere on land. At Luggiebank I found two under a corrugated sheet, bedded into the flattened nest of a field vole. Hibernation is not always continuous, and I have seen frogs ashore at my favourite quarry pond in December.

Unlike the frog, which is a nervous, volatile beast, the common toad is a placid, slow-moving creature which is easily tamed, which can be taught to come to a signal, and to take food from the hand. At Firknowe I had one that spent 100 days under a flower pot in my cold greenhouse until he departed to his spawning pond. In July I noticed a toad beside that flower pot, and he had to be the same one because he handled tame and took food from my hand with minimal coaxing. He over-wintered with me again, then disappeared for good. For the past few years I have had one under an old railway sleeper near my hen-house.

I like toads, with their warty leopard skins, shark mouths and tongues like flame-throwers, but a lot of people dislike them and won't even touch them. The warts are part of the trouble. They secrete a poisonous fluid, and the literature is full of dark innuendo about the toad's power to harm. The fact is that you can handle hundreds of toads, as I have done, without ever realising that the warts secrete anything at all. The fluid is released under stress of fear or excitement. Rescue one from a dog or a crow and you will almost certainly find it wet.

Toad spawn bears no resemblance to the jellyfish mass produced by the frog. It comes in strings which get wound round water weeds by the movements of the female during the hours of spawning. A big female will produce an egg string five to ten feet in length. Like young frogs the young toads end up in odd places after they leave the water, and at this stage are preyed upon by many enemies. Adult toads are not tasty bites for most predators, but by the time the predator finds that out the toad is dead. Rats will skin them and eat the insides. Crows and ravens open them along the belly and eat only the insides.

Because frogs and toads refuse to be moved from their spawning ponds the way to stock an unused pond, or a newly made one, is to place spawn in them. Or tadpoles. I have introduced a lot of tadpoles to one pond, a traditional one, where the frogs are massacred every other year by youths with air rifles, and the spawn dragged ashore where it soon dries out to useless tissue.

Frog and toad tadpoles have to go through a long metamorphosis before they begin to look like their parents, whereas the legless tadpoles of the newts look like theirs from the beginning. Scotland has the same three species as England and Wales — the warty newt, the smooth newt and the palmate newt. Warty indeed is the warty newt, but in Scotland it is more widely known as the great crested newt. It is the biggest newt in Europe, the palmate being the smallest. Male warty newts up to 145 mm in overall length, and females up to 156 mm, have been recorded at Dunbar, which is close on the lengths of smooth and palmate newts put together.

Like the toad the warty newt secretes an irritant fluid which probably is a deterrent to predators. You can easily squeeze some out with your fingers, and taste it if you like, as the late Dr Malcolm Smith did. He wrote: 'It has a bitter taste on the tongue at first, but this soon disappears and is then followed by a burning and smarting sensation in the whole mouth which persists for a considerable time.' I have always taken his word for it. It takes five or six months for the newt to become terrestrial, and once ashore it stays there until it is ready to breed, which may be three or four years later. After the breeding season is over the newts go ashore again. Places to find them are under stones or logs, under tussocks, in cracks in the earth or rock crevices.

The smooth newt and the palmate, especially the females, are very like each other. The male palmate in spring breeding dress has a fine black filament projecting from his blunt-ended tail. His dorsal crest is wavy, not straight edged as in the smooth newt. The smooth newt is a low ground species; the palmate is the newt of the hill country and on high ground in the Scottish Highlands will be the only one. Smooth newts leave the water after egg laying and only rarely hibernate there. On the high ground of Scotland, where shallow pools and puddles are likely to dry out, palmate newts may be found in the water at any time.

I introduced all three species to Palacerigg Country Park between 1972 and 1975.

# HARES AND RABBITS

A fox bringing food to the den.

Long before foxhunting became the fashion the brown hare was the beast of the chase, one Royal Duke describing it as 'the most marvellous beast that is'. It is still hunted — in some parts of Britain by beagles and harriers, at coursing meets, and by lone rangers using a greyhound to catch one for the pot. Then there is the rare character who uses his bare hands to catch her — lifting her from her seat (form).

Using the feminine gender for the hare was the custom in olden times and there are those who still use it. The ancients had a lot of trouble with gender, seemingly unable to make up their minds whether it was he, or she, or both. They thought it was a magical beast. Some thought that bucks could give birth; others that it changed sex by the month. Pliny thought it was hermaphrodite, reproducing equally well without the male. The belief that witches could turn into hares was widely held and I heard about it in the Cantabrian Mountains, in Spain, in 1928. An intriguing beast, to be sure.

Looking a big hare in the face sets fancy free and, when I let mine roam, I get the impression of a sly, crafty animal with goat eyes and a split-lipped, buck-toothed grin. My wife often said that she half expected our tame hare Maukin to speak to her at

*opposite.* Leveret (a young hare) well hidden in the lush grass.

133

any moment. When he did it was hare-speak, a grunt. He and I often had nose-to-nose grunting conversations. One thing all hares can do, and that is scream like a child when wounded or frightened.

Leverets are easy to rear, but their temperaments vary when adult, and some of them will bite. Our Maukin never bit anybody, but he was wary of strangers and would kick if handled. He used to kick the dogs when playing with them, sometimes forcing a low growl of warning from the terrier who didn't mind his walking over her when she was in her basket. He often slept with her, or my big German Shepherd, or both. When both were in the basket he slept on the German Shepherd's back.

I reared a number of leverets over the years, brought to me by people who had found them 'abandoned'. Those that hadn't been away too long I returned to the place where they had been found. The others I allowed to go or stay around as they wished. In later years I put such leverets in with the roe deer at Palacerigg Country Park until they were fully grown and could be allowed to go. Before they were turned loose I was sometimes able to arrange for a strange dog to run up and down the outside of the deer fence, barking at them to put them on their toes. My own dogs would have made friends with them.

Brown hares feed mainly from dusk onwards, and lie up during the day unless put on foot, like the one my dog stepped on in her seat. In quiet places like where I live they are more likely to be seen on foot late in the morning or early in the evening. Yet they attract attention to themselves in full daylight during their courtship displays, year after year on the same ground. At this trysting place the hares box and kick, and chase each other, and it is not uncommon to see a buck following a doe along a hedgerow or across a road, or across a field into a wood.

Spring is the peak period for leverets, but many are born before the madness of March and during the following six months. My wife and I tracked a hare in snow, linking up the breaks in her trail, to a tree beneath which she had given birth to her leverets. Both were dead. A hare shot in the last week of September was brought to me for examination because she had four furred leverets inside her ready to be born and two others in the early stage of another pregnancy. This is the only case of superfoetation I have come across.

Superfoetation is to become pregnant while in an advanced state of pregnancy. Pliny knew, and wrote, about it nearly 2000 years ago: 'The hare . . . practiseth superfoetation, rearing one leveret while at the same time carrying in the womb another clothed with hair, and another bald, and another still in embryo.' Pliny's great work contains mountains of fancy, but he was right about the hare. Nobody believed it until scientists proved him right 1800 years later. Superfoetation is rare in this country. The hare is notable for another phenomenon — refection. During the day it swallows its soft droppings, which pass through it again, to become voided a second time as dry pellets drained of nutrients. You can easily observe this in a tame hare.

The brown hare doesn't burrow. It makes its sleeping nest (form or seat) in a grass tussock, or among threshes, or in woodland. The leverets are born in such nests but quickly separate into forms of their own. They are born fully furred, with their eyes open, and are able to move about soon after birth. In their early days the leverets may fall a prey to men, dogs, foxes, stoats, crows, buzzard, owls or harriers. When grown up their main enemies are man, greyhounds, eagles, foxes and wildcats.

Because of its long hindlegs the hare is at its best when running uphill, which is the way it will run if it can. Downhill it will run a diagonal. It has its favourite exits from fields and woods, so is easily netted by poachers who know its routine.

From time to time I'm asked if I can pick a hare out of its seat, and when I say I can the reaction is usually one of surprise, until I admit that the six or so I have picked up were mountain hares, not brown ones. I've seen only one man do it. The late Brian Vesey-Fitzgerald said he could do it — almost. I can't do it, even almost, perhaps because I never really worked at it. It is a neat trick with the brown hare; much easier with the mountain one.

The mountain hare (blue hare, varying hare, Arctic hare, Irish hare) is smaller than the brown and about three-quarters of its weight. Like the stoat it moults to white in winter, but not always completely. Unlike the brown hare it digs burrows in peat or snow, and will regularly be seen sitting at the entrance, looking out. Although it will usually run when a person approaches too closely, it will sometimes flatten down. When it is squatting among stones or heather it may run, or crouch, to be knocked over by a stick or to be picked up by hand. The half-dozen or so young ones I have picked up made a rumbling sound in their chests. Full adults scream like the brown hare. Ray Hewson, who has handled a lot of mountain hares, noted a faint growling sound from them.

During the day the mountain hare usually lies up at the mouth of its burrow, or in heather, or among rocks. It feeds downhill at dusk and uphill at dawn, but it is highly sensitive to changes in weather and can be seen in daylight when there is a threat of rain. Over the years I have seen them feeding by day on the Lecht during a light snowfall, hares white and not so much white on dark ground beginning to freckle with the settling flakes.

Early in 1951, when the winter was savage, I saw hundreds of white hares in a Perthshire glen, by the roadside and hopping round the drifts, and sometimes bumping into my car. Blizzards force them to shift ground and when they come down they might be seen mixing with brown hares on low moors and fields.

My most memorable sight of such a movement happened below Schiehallion when I was driving with a police officer to Bridge of Gaur to see the stalker at the Barracks. The light was going fast, and then the snow began to fall, swirling against the windscreen, so that it became difficult to see ahead, with or without headlamps. It came down in great goose-feather flakes, playing tricks with our eyes, and when the first white shape appeared on the road it might have been some witchery of the lamp-lit snow-dance. But it was a mountain hare.

And then there were others, in twos and threes, some hopping, some scampering — white, spectral shapes flitting silently through the headlamp beams before vanishing into the storm. That night there was a blizzard and by morning the hills were white as far as the eyes could see. And the white hares were down on the road.

The mountain hare feeds a lot of mouths. Depending on its size it is taken by eagles, foxes, wildcats, stoats, buzzards, ravens and crows. At certain times of the year up to half the fox's prey may be mountain hares. I worked with a pair of eagles that had over 80 in 12 weeks, and the eaglets were fed little else during their 11 weeks and five days in the eyrie. On high moorland roads motor traffic takes its toll, providing food for the scavengers and pick-ups for the like of me to take home for my wife's wolves.

Jokes about the rabbit, especially about their fecundity, are legion. Jim Lockie and I became the subjects of a joke when I was helping him with an eagle survey on Lewis in 1955, mainly to find out the extent of predation on lambs. We decided to replace any lamb found in an eyrie with an equivalent weight in rabbit, so that the eaglets wouldn't go hungry or the cock eagle be pressured to extra hunting. I arranged with a rabbit catcher in Perthshire to send us a supply of intact dead rabbits to Stornoway, which were then stored for us by a local butcher in his chill. In the event we found no lambs in eyrie, so didn't need the rabbits, which were high when they went out with other offal. And the rumour was born that Lockie and Stephen had intended bringing myxomatosis to Lewis. Someone did eventually ship live, diseased rabbits to the island, and the old rumour became a joke.

Myxomatosis almost wiped out the rabbit population over most of Britain, the operative word being almost. A 98 per cent die-off was a lot of rabbits, but the surviving 2 per cent soon hefted themselves, and populations increased slowly despite recurrent outbreaks of the disease. Some parts of Scotland were never touched, and rabbit catchers remained in business. During my time at Palacerigg I was able to buy 200 rabbits a year from one source, in two batches of 100, to be deep frozen for the Park carnivores. In 1988 the disease struck some neighbourhood warrens, while the rabbits at others remained clean. The rabbit is coming back, despite shooting, snaring and gassing; despite predators; and despite myxomatosis. How far it will succeed in hefting itself remains to be seen.

When I was at Firknowe I built a rabbit warren in the spacious corner-of-the-garden-wall outhouse, with an exit into a netted enclosure in the spinney, and installed a pair of wild rabbits, which I was able to view, above ground and under their earth canopy, from the big, south-facing window. If they bred, and kept breeding, I could skim the surplus for my stoats and weasels. That was one reason for keeping them. I wasn't very clear in my mind what else I would do with them, except watch them and wonder about myxomatosis. When the disease struck at two neighbourhood colonies my rabbits remained clean, despite some contact with outsiders against the wire netting at night.

My terrier killed a number of diseased rabbits, one of them in the driveway to the house, and within a week chased two into an old drystane dyke. I recovered these, both clean, and put them in the cold greenhouse. Myxomatosis was striking here and there, but not everywhere at the same time. I put the new pair with my own for 48 hours. The old buck was rough with them, chasing them out to the netted enclosure, but during the day they all slept together in the big chamber. I took the pair out, and they remained clean. So did mine, and right through the next spring and summer when I stopped breeding them.

But the idea of having a tame colony of wild rabbits was always at the back of my mind. I hadn't the space for such a venture, and nearly ten years were to pass before I had — at Palacerigg. An area of nearly a fifth of an acre was netted off — a grass area with a row of mature hardwood trees at one end — and secured against foxes and cats. Two hoppers were installed to hold feeding pellets. My wife reared, in succession, two lots of kits, the first from a nest I found in the spinney and the second from one a friend had found in her rhubarb patch. That gave me a colony of six, completely tame and all clear of fleas.

The colony of tame wild rabbits soon became a great attraction, and school children could watch rabbits, big and small, popping in and out of burrows in daylight. They could watch a master buck in the north-east corner driving off rabbits of lesser status, or thumping the ground with a hindfoot when alarmed. The netting was a neutral area, so the children were able to feed rabbits of all sizes there, from masters to beasts of low degree. Myxomatosis came and went, but the enclosed colony remained clean.

In a rabbit warren there are group hierarchies of master bucks and their does. These does have their young in a nest in the warren. Does of lower status have to move out and have theirs in a temporary burrow known as a stop. The doe makes her nest of grass, to which she adds wool plucked from own body before her young are born. She feeds them at night, visiting them only once.

Superfoetation is unknown in the rabbit, despite Pliny's assertion to the contrary. Like the hares it refects, swallowing its daytime droppings for redigesting. When rabbit numbers are high, many litters conceived are never born; they are reabsorbed in the uterus. In dense populations this pre-natal mortality can be as high as 60 per cent. According to the Old Testament the rabbit is unclean because 'he cheweth the cud but divideth not

*opposite.* A Short-eared Owl which returned to Palacerigg after several years absence after David Stephen had improved the habitat.

the hoof'. Whoever wrote Leviticus must have known about refection. The pig is unclean, but the other way round: it divides the hoof, but cheweth not the cud.

# WILDCATS AND FOXES

Young fox playing with Lisa.

The stereotype of the Scottish wildcat is a bristling, snarling fury, with its ears flattened against its skull and teeth bared to the gums. That is the stereotype because it is the way most photographs and drawings show it. That is the way most photographs show it because most photographs are of captive wildcats, taken wild, or of animals at bay or cornered. The picture is false. Of course wildcats can be like that. But the wildcat going about its everyday affairs has its ears up, its mouth shut and its teeth locked away until needed.

The wildcat was reputedly untameable, and many people still believe this. Untameable it is if you wait until its eyes have been open for some time before you go to work on it. Tameable it is if you take it before its eyes have opened. I know because my wife reared one such on the bottle, and he is still, fifteen years on, the same gentle and genteel beast he was when cuddled in her lap as a kitten. My experience with a hybrid was rather different. I got him when he was about six weeks old, and it took me seven weeks of hard work to tame him, at the cost of some tooth-and-claw surgery.

The mate of my tame male wildcat came to me when her eyes had been open for a few days and, although she was reared on the bottle, side-by-side with him in the same lap, remained hostile and untouchable to the day of her death. I had to wear a

*opposite*. The Stephen's Wildcat kitten at Palacerigg.

141

David Stephen fondling
Teuchter the tame wildcat
at Palacerigg.

fencing mask and horsehide gloves when I was taking out her
kittens, and she tattooed my right hand for life with her tusks.
Anyone wishing to hear the sing-song, wailing pibroch of a
wildcat fuming had only to face her at too close range. Himself
didn't know the music, or if he did he never sang to it.

Persecution by keepers throughout the nineteenth century and
up to the outbreak of war in 1914 hit the wildcat hard, leaving
it an endangered species south of the Great Glen and thin on the
ground north of it. It was driven to seek refuge in the wildest
places on the hill, and survived. The period during the war gave
it a respite and in the post-war period it began to spread. The

Second World War gave it a further respite; again it hefted its numbers and spread even farther, as far south as Stirlingshire. On Forestry Commission ground there I came upon a wildcat and watched two kittens at play on a later date. In the Loch Lomond area I watched a female, on several mornings, returning to her den in a corrie, above which I was photographing golden eagles.

It has long been known that wildcats and feral domestic cats sometimes breed, producing hybrids. In 1957 I tamed one such. He looked like the real thing except for his long, and tapering, tale. The tail of the wildcat is blunt-ended, ringed with black, and black at the tip. My view of wild crossing with domestic is that it has to be wild tom to domestic female. I can imagine a wild tom accepting a domestic female in oestrus, but I can't imagine his doing other than driving a domestic tom right off his territory, any more than I can image a female wildcat going a-wooing round the houses in the glen. All this is speculation.

If hybrids between the wildcat and the domestic cat are sterile it follows that the Scottish wildcat is pure blooded. But not all Scottish wildcats are, which means that the hybrids are fertile. What, then, is the extent of the dilution? Professor Suminski of Poland did a lot of work on this problem and I asked him his opinion when he stayed with me for two nights some years ago. Polish wildcats came top at seventy-three per cent pure. The Scottish wildcat came out at sixty-six per cent pure against a European average of sixty-three per cent.

The wildcat isn't often seen on foot by day, unless you happen to be in the right place at sunset or sunrise. Most are seen dead in traps. Others are killed on roads. I picked up two one winter on Schiehallion. I think it would be true to say that any big cat, tabby or grizzle, seen on the hill at night, or caught in the headlamps of a car on a Highland road, is as likely to be a wildcat as a domestic one. The wildcat is now a protected species.

I have often said that if I were a fox in certain parts of Scotland, facing the daily risk of being trapped, snared, shot, gassed or poisoned, I would flit to one of the English shires and take my chance running before foxhounds. Some years ago I changed my mind. Now I would move into town, where road traffic would be the greatest hazard.

Nowadays, townspeople see foxes as often as, perhaps more often than, the country dweller, and the urbanite is indisputably more tolerant of them, although this is gradually changing. The average countryman, out with a gun and seeing a fox, is more likely to shoot it than to leave it alone. But the producers of eggs and table fowls are no longer interested in the fox as a predator, because their birds are either in battery cages or broiler houses. Sheepmen and the rearers of game birds still kill foxes as routine, many of them using illegal gin traps and poison. More and more country people — a trickle, not a spate — put out food for foxes for the pleasure of watching them from a window. My daughter has foxes and badgers outside her kitchen window every night, every year, from September to April.

Urban foxes don't have to work very hard for a living because many people put out food for them, or for stray cats, and a lot of them buy food specially for both. Human wastefulness with food helps, and the fox is a first-class scavenger. I watched one at first light outside a chicken bar,

*opposite.* Portrait of vixen. Rabbits and mice are pretty safe near the den; the fox goes further afield to hunt.

Vixen (female fox) in gin trap. These traps are illegal now but wire snares are not and are just as inhumane and unselective.

wading through the litter looking for left-overs, then crossing the road to drive a rook away from a half-eaten fish supper. The matron of an Edinburgh hospital sent me a photograph of a badger she fed every night outside her sitting room window.

From west to east across Scotland's central belt I have spoken to townspeople about urban foxes. Most people liked seeing them; a few had them in the garden; a tiny minority wanted rid of them. One woman had a vixen with cubs under her garden shed and was concerned about their safety. Another person in the same situation wanted to know how to get rid of them. Several people whose cats had disappeared wondered if the neighbourhood fox might have taken them. Possibly. I've seen a fox run a cat into a tree, and watched another one killing a young cat from my neighbour's farm. But I have also seen a fox skulking round a fizzing cat in a farm stackyard, and I have never seen a dead cat at a fox den.

Foxes have a reputation for being cunning. They are certainly very wary. During a winter fox drive one beast, wounded in a foot, ran the top of a drystane dyke for more than a hundred yards. That got its scent off the ground but didn't fool anybody because the blood-stained tracks were clearly printed on the snow. At another fox drive I watched a fox climb into a tree, and said nothing to the beaters. One terrier had her suspicions but her owner called her in, thinking she was just wasting time fussing me.

Some foxes are habitual tree climbers. One dog fox spent long spells, during the day, in a tree near the den while the vixen stayed on the ground. My own tame dog fox, Glen, spent some time, almost every day, throughout the year, fourteen feet up in a tree. One visitor, looking at him, asked me if he was a stuffed fox. I reckon that Glen must have been the most photographed fox in the country. Fiddich, his mate, never left the ground.

Back in the 1950s I reared three fox cubs, with the vague notion of being able to keep contact with them when they went back to the free life. The idea didn't work out, because they had learned to trust people and two of them were shot within a week. Some excellent research on wolverines was done in Sweden in this way. David Macdonald perfected the method and spent fifteen years in the field, using every modern technique, to come up in the end with his classic *Running with the Fox* — a book that sheds a brilliant light on many aspects of fox behaviour not even guessed at before.

The late Mortimer Batten coined the term bachelor fox for one carrying food to the den of a mated pair. Years later I used it to describe the same situation. That much was known. Two vixens birthing in the same den suggested some sort of socialising by four animals. I have seen four foxes feeding off a dead deer in winter, and two following a sickly roe deer that was ready to collapse. David Macdonald has shown that the fox's social life is much more complex than anyone imagined and his book is set reading for anyone interested in foxes.

Like leverets, deer, fawns and owlets, fox cubs suffer from rescuers of abandoned babies. Vixens do not desert their cubs, and the dog fox will rear them if she is dead and they are past the milk stage. With the help, maybe, of a bachelor fox? Fox cubs are easy to rear and become quite tame, absurdly so in many cases, but should be left in the wild unless one has a good reason to take them from it. Hand-reared foxes are on offer every year but find few takers.

Tarf the family labrador with Spider the tawny owlet.

# RODENTS

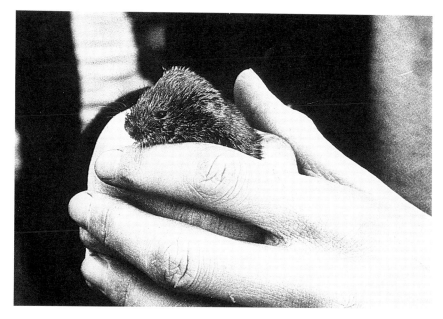

A bank vole.

The brown rat is resourceful, cunning and aggressive, bold and shy, wary and reckless, clean and dirty, wasteful and provident, highly adaptable and hide-bound, quick to learn and slow to forget. I learned that it was all these things after I moved to my farmhouse at Luggiebank in February 1949. There I found that I had inherited a rat colony, and I decided on a two-pronged approach to them, to study them while exterminating them.

While the terrier and the cats carried out executions, and I did some live-trapping and marking, I made the stable a no-go area where I could do a bit of rat-watching in peace. I baited them to a heap of oats and poultry pellets in the centre of the stable floor, and left a red light on all night. Rats don't like bright light at night, but the stable lot didn't mind the red one, and if they wanted the goodies they had to come to the middle of the floor instead of skulking around rubbing shoulders with the walls. Once boldness had taken over from wariness I boxed myself into a stall and tried them with eggs.

You'll know the old story about rats and eggs. It was, and is,

*opposite*. Bank Vole, or red-backed vole, so called because of the rich chestnut colour of the upper back.

that when a rat wants to move an egg it finds a partner to help. One rat then lies on its back, clasping the egg, while the other takes it by the tail and pulls it along to wherever they want to go. I never gave the story a moment's belief, and I have never met a believer who had actually seen it. One night I killed it stone dead in the presence of believers. All my stable rats ever did with an egg (watched by faces at the window and me in my loose box) was nose it across the floor into a hole, even hefting it a few inches if the hole wasn't at floor level.

Rats are wary of any new object in their path. When I put an upturned bucket, with an egg on top, beside the heap of grain and pellets, they walked and scurried round the stable, rubbing the walls, while they studied it. They moved up to it, then backed off, several times, before padding round it to pick up a pellet or an icker of corn. I left the bucket there for a week, then removed it. That night they followed the path they had taken round it, and made several trips before they realised it wasn't there.

One day Jess came into the house and told me there was a hedgehog on the meal store shelf. I didn't believe her because there was no way a hedgehog could have reached a perch as high as I am tall, but there had to be something that looked like one so I went out to see. There was a beast on the shelf all right, with head down, back arched and wiry hair bristling like a hedgehog's spines. But hedgehogs don't have scaly tails, or the kind of face that turned to greet me, with buck teeth threatening. I was looking at a rat, an outsize one, and when I stepped inside its toleration distance it took the shortest way down — via my face. I sidestepped it, and it bumped my shoulder before ending up among the stacked corn and meal sacks on the floor.

That made it dog business so I brought out Nip, my Border terrier, to carry out the execution, which she duly did, getting bitten twice herself while I was moving sacks to give her elbow room. The rat weighed one and three-quarter pounds and measured eighteen and a half inches from nose to tail tip. A rat record? Not at all. Rats of two pounds and two and a quarter pounds have been recorded, but there must be few that reach such a size and weight in their short lifespan of one or two years.

The brown rat arrived in Scotland in the nineteenth century, a hundred years after it landed in England from ships. It reached the New World in ships. It conquered the world via ships. It is highly adaptable, able to live almost anywhere, from sewers and coups to mountaintops, mostly at the expense of man. It eats crops, growing and in store, it fouls foodstuffs and spreads a number of human diseases, including leptospirosis. Many years ago I knew two colliers who went down with this disease. One died. The other became mentally ill, permanently. Once, when down a coal mine, I watched the colliers feeding the rats. One threw a bit of pastry to a rat and was jokingly chastised by another who said: 'Hey! That's my rat. Don't spile it wi the fancies!'

The black rat, properly called ship rat, reached England in the eleventh century and was the one that caused the Black Death. It came to Scotland in the eighteenth century, about the time of the '45 Rising, and Jacobites blamed George I for sending it. It is a sleeker, longer-tailed rat than the brown. Some black rats are brown and some brown rats are blacks, but black rats in a warehouse or dockland buildings will almost certainly be the real thing. This species has been declining in numbers over the years, but is still present in some strength in four parts of Scotland: Clydeside, Forth, Tayside and Aberdeen. My friend Dr Jack Gibson tells me that the Clydeside population is steadily declining, seemingly because it can't hold its numbers up without immigrants.

The water vole, commonly called water rat and frequently mistaken for one, suffers as a result although the two are far from being look-alikes, in or out of the water. The confusion arises because brown rats in the country sometimes move afield for the summer months, burrowing in hedge bottoms and in burn banks, and in some cases taking over the ready-made tunnels of the water vole. Age for age, vole and rat are about same size, but the vole is dark brown, with blunt face, and a furred tail more than half its body length. When I am close to a water vole peering from its burrow I get the impression of a face like a porcupine's. In the north of Scotland water voles are almost black.

When the young water vole leaves the nest it might easily be mistaken for an adult field vole, except for the length of tail. At this stage it is well within the prey range of predators like the weasel and the tawny owl. I have seen a heron taking one of this size. Foxes, pike and mink are other predators. One year a pair of water voles took over an islet on the big pool at Palacerigg, and I saw them often, early and late in the day,

Water Vole in a hollow log; sometimes these creatures are wrongly called water rats, but the water vole is wholly vegetarian.

sometimes ferrying lengths of greenstuff to their burrows. I rarely saw them on the banks of the burn flowing into and out of the pool. They reared young on the islet and I saw four at one time, never more. Next summer I saw no voles, or sign of them, and wondered if their disappearance had anything to do with the mink that took up residence for a while on the islet. I box-trapped the mink and put it behind bars.

The following spring the islet had voles again. I saw their signs before I saw them at the latrines where they had deposited their droppings — Monty called them vole's macaroni — and bits and pieces of food. Voles also set scent from their flank glands at the latrines. The pool pair came readily to a bait of apple or raw carrot. I caught up a pair of the young and kept them for a few weeks, housed beside a small temporary pool, where I tried them with a variety of plant foods, including potato and turnip and beet. I also watched them swimming under water in my big fish tank where they were filmed. When I released them at the big pool they dived in with a *plop* and swam straight to the islet.

The field vole, blunt-faced and short-tailed, is a much-researched animal whose numbers, in certain habitats, fluctuate on a cycle of just under or over four years, sometimes breaking away at the peak to 'plague' density. Such 'plagues' have been recorded since biblical times. The late Aristotle gave a description of a plague of mice that could be applied, without editing, to a vole cycle today. In the years 1875/6 and 1891/2 there were such 'plagues' in the Scottish Border country. The voles ate out the grass and the sheep had to be hand fed throughout the summer. In 1953 there was such a breakaway in the Carron Valley Forest in Stirlingshire.

Young plantations, from which other grazing animals are excluded, make ideal habitats for voles because of the abundance of grass as food and cover, and it is in such habitats that their cycles are most obvious. The number of voles per acre can be as low as five–ten, or as many as 600, depending on the stage of the cycle. In the Carron Valley Forest breakaway the voles did more than eat down the grass; they destroyed a whole plantation of three-years-old Scots pines and, although not great climbers, managed to bark some older trees up to twelve feet from the ground.

Nigel Charles and Jim Lockie, neither of whom I had then met, studied this 'plague' in detail. It is a characteristic of such breakaways that when the vole numbers go up the predators move in. Jim Lockie told me later that before the breakaway there were ten pairs of short-eared owls in the forest, holding about 400 acres of territory per pair. At the peak there were forty pairs holding forty five acres per pair. There was also an increase in the number of stoats, weasels and buzzards. Everything that ate voles fed fat on them and the owls reared bigger families. It was the increase in the predator force that attracted me to the forest, and Jess and I visited the area as often as we could to see owls, buzzards and kestrels hunting or perched along the roadside stuffed with voles. Sightings of stoats and weasels were frequent. Later I went with Jim Lockie round his traps, when he was weighing, sexing and marking both species.

Later on, when I was keeping stoats and weasels, I bred some field voles under semi-natural conditions. Two families, separated by fine mesh wire, behaved aggressively towards each other through it. A mother, disturbed too soon after birthing, will sometimes eat her young. I found that leaving them alone was the best policy.

The bank vole is a mouse-size rodent, with chestnut-brown upper fur and grey or buff underfur. Its tail is half its body length, dark above and pale below, and often carried gay. Like other voles it has a blunt face, but it is more mouse-like in profile than the field vole, as well as slimmer and sprightlier. This species prefers deciduous woodland, or scrub, with plenty of ground cover, but it also likes hedgerows, and banks thickly overgrown with brambles, dogrose, or other trailing and creeping plants. Once found, it can easily be watched by day. Besides the leaves and steams of plants it eats berries, seeds, fungi and roots. It also takes insects and their larvae. Orkney has a vole of its own.

Men have always, more or less, taken mice for granted; women have opted, more or less, to be scared of them. Not my Jess, though. She is quite happy to forget about the television and watch a handsome woodmouse stealing her chocolates. Yes, we nearly always have one in the house, and the generations seem to pass the word on because they all know the way from under the stairs into the sitting room. If a house mouse turns up I have to catch it in a Longworth live-catcher and put it outside. But the woodmouse is a guest. A dear friend of ours used to call the house mouse the ordinary mouse, while the woodmouse was the braw moose.

There is a refreshing ambivalence about our attitude to mice compared with our attitude to the rat. Wherever man is, the house mouse will try to be and it is as a commensal that it is best known. Before St Kilda was evacuated, house mice occupied houses and byres; after the evacuation the woodmice moved into the buildings and the house mice disappeared. On Lunga of the Treshnish Isles, where there are no woodmice, there were and are house mice. Eighty years after the island was evacuated they were still there, living the free life afield like their ancestors on the eastern steppes. Yet, when Fraser Darling moved into a bell tent, in 1937, to do a job of work on the island, the house mice moved in and began to share his stores.

The handsome woodmouse, with its shining, boot-button eyes, is the species Burns turned up with the plough. Although it is a mouse of the woods, fields and hedgerows, it will occasionally make a foray into a house, like our succession of visitors at Palacerigg. Mostly a woodmouse in a house is one that has escaped from the cat. When we had cats we were seldom short of an escaped woodmouse. Now that we have no cats we have our temporary guests.

The woodmouse is nocturnal, reluctant even to face bright moonlight, so it isn't an easy beast to see out of doors. At Firknowe I kept some for a while, but only one reared a family. I got a great deal of pleasure watching them climbing, using the tails as a fifth hand. Some would take food from my fingers, but

The woodmouse is a nocturnal beast. It is a main prey of Tawny owls.

always at full stretch, wary and nervous. By the way: never pick up a woodmouse by the tip of its tail, because the skin will sough off, exposing the bone. The bone dries up and breaks off, leaving the animal with a permanently shortened tail.

When I moved to Luggiebank in early 1949 there were red squirrels in the pinewoods and mixed woodlands, and I could usually persuade one to come and sit on my head to eat titbits I had placed there. Mind you, I was always in a tree hide, photographing something or other, and there was a taut roof of thick felt between my head and the squirrel's feet. At Atholl I baited a red squirrel with fruit-and-nut chocolate, and bribed it to ground-level, where I filmed it sitting up eating some of it before burying the remainder in the leaf litter.

In the early 1930s, when it was clear that the red squirrel was increasing its numbers, the owner of a small estate, which was mainly mixed woodland, introduced a pair of red squirrels. We put out trays, loaded with nuts, at head height from the ground, and fixed roomy nest boxes much higher up. The squirrels ferried the nuts to the boxes, and continued to do so each time the trays were replenished. They ate in, and on top of, the boxes but never slept in them. This pair also visited the bird tables throughout their range, draping themselves over the nut dispensers much to the annoyance of the tits.

Red squirrels lost ground during the Second World War, and disappeared from some places, including the wood where they had been introduced, but they were holding up in the woods I knew, and in the glen not far from my farmhouse. I saw my first grey squirrel there in 1953, and from then on they gradually colonised the surrounding woods and spinnies. But there was no immediate takeover, and when I moved to Firknowe in 1958 there were still red squirrels in the nearby woods. The last I saw of them was in 1963. During the next ten years I heard reports of a red squirrel having been seen in this wood or that, but I was never able to confirm any of them.

At Firknowe I resisted the temptation to add a pair of young red squirrels to my animal collection because they were, by then, obviously becoming rare. I might as well have done, because they disappeared anyway. I have been offered adults since then, but declined because I prefer to rear my own. When I moved to Palacerigg I was offered a pair of young red squirrels in exchange for two wildcat kittens, and I accepted. I parted

*opposite*. Red Squirrel in a Scots pine. Alas, these popular little animals have all but disappeared from our woodlands.

with the kittens but am still waiting for the squirrels.

Can the virtual disappearance of the red squirrel from central Scotland be attributed to the arrival of the grey? The grey has certainly replaced the red, but was it a take-over by force? The explanation is likely to be far more complex than that. The red squirrel has had its ups and downs before. In the early eighteenth century it became extinct in southern Scotland, and almost so in the Highlands. It was reintroduced from England at various times between 1772 and 1872. Very likely others were brought in from Scandinavia as well. The introduced animals multiplied and spread, reaching such a peak of numbers at the beginning of this century that they destroyed 100,000 trees in one year at Glen Tanar. This peak was followed by another decline, and that was two decades before the grey squirrel arrived from Canada.

Both are true tree squirrels, the red especially spending most of its life there. Surprised on the ground it will always go up. I have seen one go up a telephone pole, and another go up the ivied gable of a house. Barrett-Hamilton has recorded how a red squirrel, confronted by a Highlander and his dog on a treeless moor, climbed the man to escape from the dog.

Like the red squirrel the grey can easily be brought to a bait, and in public parks will take food from people's hands. Like the red again, it will visit bird-tables and ferry food away. One I watched regularly used to wrap itself round the nut dispenser (like the red before it) making it almost impossible for the tits to feed. One great tit solved the problem by alighting on the squirrel's rump and feeding from there.

Neither squirrel hibernates; both have to feed regularly throughout the winter, or die. I have often watched Highland reds feeding in larch trees on sunny winter days, and not so sunny ones, when the snow was lying deep and the ground frozen.

Red Squirrels eat the seeds
of young pine shoots; but
they like chocolate and
biscuits which they will
take to some hidy hole and
come back for more.

# NIGHT FLYERS

A Woodcock returning to its nest in wet woodland. The male flies the same circuit around his territory every evening at dusk.

When I lived in my farmhouse there were pipistrelle bats in the loft, and grand pleased I was to have them there. On fine nights I could sit outside the front porch and listen to their small-talk as they wavered up and down, through the trees and round the house. They are very talkative on the wing and, when there are many of them together, they can make almost as much noise as swallows, although the human ear, with age, gets past the stage of hearing them.

Pipistrelle eats most of its prey, gnats and other small insects, on the wing, but larger ones are quite often carried to some favourite perch to be dealt with, just as the sparrowhawk uses its so-called plucking stool. In midsummer, bat-light for pipistrelle is sunset, and it will be hawking around before the owls are on the wing; in spring and autumn it may come out at the same time, but it is more likely to wait for owl-light. When drinking, it will sometimes settle on the edge of a pool and sip; at other times it will drink on the wing, hovering like a kestrel just clear of the water. It will drink from bird-baths in gardens, and I have watched one paddling in mine — a deliberate act, not an accident.

The flying season is from March to October, but it isn't

*opposite*. A Tawny owl disturbed in daylight in autumn.

161

uncommon to see pipistrelles taking a turn around on sunny days at other times of the year. I watched three beside Loch Tulla on a February day, catching midges under the trees, when Rannoch was blanketed with snow and the drifts were six feet deep. During hibernation, as in the flying season, pipistrelles are gregarious, and large numbers can be found together, although these days there are not so many of them around as there used to be. They mix readily, and amicably, with other species. They hibernate in buildings, caves, holes in walls and in trees, and can squeeze into crannies seemingly too small for a winged creature of the size. I once found two in the nesting burrow of sand martins. Pipistrelle runs neck and neck with the pygmy shrew for the title of Britain's smallest mammal.

Daubenton's bat is often called the water bat, and is the one most often reported taking the fisherman's fly. Its wingspan is two inches or so greater than pipistrelle's, which isn't much help with identification if you are watching bats in flight. The fact that it hunts a lot over water is no guide because pipistrelle does the same. Daubenton's bat is found as far north as Inverness. In summer it roosts in hollow trees and buildings; in winter it will use caves as well during hibernation.

The long-eared bat is aptly named and easily recognised because its ears are three-quarters the length of its head and body. It likes woodlands and buildings in sheltered areas and isn't likely to be found on treeless terrain. It roosts and hibernates in trees and buildings, and sometimes in caves, especially in very cold weather. It catches insects on foliage as well as in flight, and will carry big prey to a favourite perch to be dismembered.

Natterer's bat is found as far north as central Scotland, in woodland and parkland. It catches most of its food, including moths, on the wing, but it also takes insects from foliage. It roosts in hollow trees, buildings and caves.

Bats mate in autumn and winter, and the females store the sperm until May, when they ovulate and become pregnant. They give birth to a single young in late June or in July, and the females of many species, perhaps all species form nursery groups away from adult males. They don't usually carry their young in flight, but will do so if disturbed at roost.

A lot of people are afraid of bats (the Dracula syndrome?) but most seem to be neutral. Some like them and look after them, while others hate them as much as some gamekeepers hate birds with hooked beaks. I tried in vain for thirteen years to attract bats to Firknowe, but they preferred the nearby farm buildings. I see a few near my house at Palacerigg but they don't live with me.

Almost every year I get queries about bats, usually from people who want rid of them. One woman said bats had moved into her loft and were making a mess with their dung. I told her to shovel the stuff up and throw it on the compost heap. Would I move the bats for her? No; I wouldn't. She was not amused. The secretary of a senior secondary school phoned to say that bats had taken up residence and the headmaster wanted rid of them. So much for education! On the other hand it is always a pleasure when I hear from someone who wants to know how to attract bats, or who has just had some immigrants and wants to know what *not* to do.

All bats are fully protected by law. The Wildlife & Countryside Act of 1981 makes it an offence to kill, injure, or handle, any bat, to evict bats from a roost or to seal off their access to it. Where work has to be done on a house or building, and is likely to disturb bats, the Nature Conservancy Council should be notified and their advice asked for.

Ask people a question about what flies at night and most will answer owls before they add bats. Yet the short-eared owl is a day-bird, mostly. It nests on the ground and catches most of its prey there, mainly small mammals but occasionally small birds. The field vole is the commonest prey item, and where there are plenty of them the short-eared owl will surely come to see, perhaps to stay. Although it is usually a silent bird it can bark and yap when its nest is threatened. Its hoot is a low-pitched *boo-boo-boo*. The owlets leave the nest any time between the ages of twelve and seventeen days but they are between three and four weeks old before they can fly. The biggest brood reared on Palacerigg was six, but some birds will rear up to twelve in a vole year. Like other owlets the young snap their beaks when threatened, and defend themselves with their feet. An owlet brought up by my wife stayed with us until the following spring, and used to fly to her hand when she held up a vole. For many years the short-eared was no more than a visitor to Palacerigg, but once we had acres of young trees planted, and plenty of grass, two pairs moved in and stayed to breed.

There is a wood not far from my house where I used to spend

many nights every year with nesting tawny owls, sometimes up beside them on a platform in a tree, sometimes on the ground where I could listen to them and watch badgers at the same time. At nesting time this owl can be dangerous, especially after dusk. Some birds will attack in daylight, and some before their eggs have hatched. Some will attack even when their young are in the trees, and flying. I quickly got around to wearing eye-shields and a flying helmet when climbing to nests around dusk. I gave more than a few millilitres of blood to tawny claws over thirty years. One bird gave me a thick ear. Another knocked my eye-shields awry and hooked the lower lid of my left eye, which is my seeing one. The eye itself wasn't damaged. Bird photographer Eric Kosking lost an eye in this way. One night I was watching badgers with the head stalker on Atholl when we were back-slapped by a tawny owl, whose young were flying and whose nest was nowhere near. A bird at Bridge of Gaur was a day attacker, and came at the head stalker and me with a ringing war-whoop every time we approached her tree.

This is a vocal owl, with a fair repertoire of owl-speak: the long-drawn-out *hoo-hoo-hoo-hooooo*, the ringing *kee-wick, kee-wick, kee-wick*, the lower-pitched *wee-wicking* of a hen brooding owlets, her sharper version when the cock calls her off for a prey, the twitting and chirping of flying young, and the cheetering of small young in the nest. When the cock hoots his signing on signal at dusk the hen will usually *kee-wick* to him from the nest, and sometimes this becomes a duet. When she flies off to accept a prey the pair sometimes fly around *kee-wicking* to each other for perhaps a quarter of a minute before she comes back to the nest to feed her owlets, passing the food to them under her body.

The tawny owl nests in holes in trees (actually it makes no nest) or in the old nests of other species: crows, sparrowhawks, herons and magpies. For several seasons I patched up one nest at which I had a hide, and the birds continued to use it. Later on I built the nests myself in places that suited me and made no difference to the birds. When an old hollow tree, long known for its nesting owls, fell into the burn, I put a barrel up in the nearest one, and the owls used it from then on. Nest boxes for owls are now available.

This owl is mainly, but not strictly, nocturnal. When it is feeding young it will hunt from before sunset until sunrise, and by day in woodland if it has to. A bird treed all night by heavy

A Tawny owlet perched on a stump during the day.

rain will certainly do a tour in a wood by day. After a night of poor hunting it is likely to do the same. One of the tawny owlets I reared was called Spider, and he was as active by day as he was by night. He used to perch above the two cats when they were hunting, but especially the big cat Cream Puff, who was a great vole hunter. When the cat was playing with a vole, as he often did before eating it, giving it a yard start before pouncing again, that was when he lost it. Spider was down, and up, and away with the vole in his talons. The cat never learned, and the owl often profited.

In May 1953 I left a nest after the cock bird had gone to roost, then came back about noon to take some daylight photographs. There was a prey on the nest that hadn't been there when I left, and I assumed that the cock had been hunting while I was away, until I found out that he often went to roost with a prey, which he brought to the nest later on in the morning. I also saw him holding a rabbit at roost all day, which he brought to the nest before leaving to hunt that night. That bird had only one eye and I often wondered how he had lost it.

I never expected that I would ever meet another one-eyed owl, but I did, twenty seven years later. This time the bird was a hen long-eared owl and, as in the case of the tawny, I would never have known about her one-eyedness if I hadn't put a hide up in a tree beside the nest. When I first looked at her brooding her five small owlets I thought she was merely keeping one eye closed, as owls often do, until I realised the open eye was the only one she had.

Long eared owls at nest in pine shelter belt at Palacerigg. Their prey is woodmice.

The cock bird was hunting over about 50 acres, half grass and half young plantation, and was bringing in voles about every 20 minutes or so during most of the night. When all the owlets were able to swallow whole prey, One-Eye began to hunt as well. And she answered for me a question I had often asked myself: how did the loss of an eye affect the bird's hunting, if at all? I got the answer when the pair began to hunt the open area of voles before the sun had set, sometimes half an hour before.

And they the most nocturnal of our owls! From my seat I could watch every strike, as often as not without using binoculars. One-Eye was killing as many voles as her mate, and sometimes they arrived at the nest within a few seconds of each other. One evening the light was so good that I was able to run some colour cine film of the pair at the nest. They reared all five owlets.

That was a vole year, and no bird prey was brought to the nest on any of the nights I spent beside it, which made them an exceptional pair again. All the other long-eared owls I had sat up with over many years brought in birds at some time — skylark, reed bunting, snipe, greenfinch, chaffinch, jay, blackbird and yellowhammer being recognisable. One cock owl, whose nest was near a farm, brought in seven young rats and a skylark before two o'clock in the morning.

I had seen broods of five before One-Eye's but not all of them survived. If food was short, and the smallest owlet died, it became prey and was eaten. One year Monty and I helped out a pair with five young by supplying them with mice and voles and bits of chicken and rabbit. All five survived.

Time was when I could look in on seven pairs of barn owls without having to walk ten thousand miles; nowadays there isn't a pair in any of the seven places. This isn't because some of the places no longer exist, because they are also absent from those that do. One far doocot, a nesting place for as far back as I can remember, has now been sealed off. Farmyards are not what they were. Grain is combine-harvested and grain stacks in farm stackyards are things of the past, so the mouse hundreds that live in them are also things of the past, and the barn owls no longer have easy prey ready to hand. There will be other causes involved in the barn owl's decline, but absence of stackyards and alterations to farm buildings have to be two of them.

When I lived in Luggiebank I was deeply involved with tawny and long-eared owls, but had done little more than watch them flying into the buildings where they were nesting. In 1950 I decided to look in on a pair nesting at a neighbour's farm only 10 minutes' walk away. I had often sat at the gable, below the entrances to the doocot, watching the birds coming and going, and on still nights I could hear the snores and whistles of the owlets framed in them. I thought the doocot would be a pleasant, comfortable place to work in, and the farmer agreed at once to give me access.

Unfortunately, it could be reached only through a trap door in a bedroom ceiling. The farm folk didn't mind the idea of my appearing in the spare bedroom in the middle of the night, although I thought it a bit much to expect of them! However, I decided I would go up and have a look at the owlets. The doocot wasn't floored, and when I stumbled between two joists my feet went right through the ceiling, knocking down a lot of

whitewash and plaster. That embarrassment decided me to watch this pair of owls from the outside, although the farmer offered me a wooden door as a floor to work on. Having the ceiling fixed was no problem.

One pair of barn owls nested every year in the chimney of a derelict steading, on a solid base of twigs piled in years before by jackdaws and added to by owlet pellets and other detritus. It was impossible to see the birds at the nest, so I could never be sure what prey the cock was bringing in, but I spent many a happy hour there at night, seated in the old fire place, listening to the hisses, snores, screeches, yaks and chirrups of parents and young. It always gave me a thrill to see the ghost-owl flickering over wall of the roofless steading, or through the glassless window frame, like a great moth, the whiteness of him plain as a badger's face in the dark, and I knowing he couldn't see me because I was sitting motionless and wrapped in shadow darker than the night.

Another pair moved into a hollow tree when they lost their place at the farm, but they still hunted the stackyard. One frosty night, when the ground had a scowder of snow, one of them flew down in front of me and proceeded to hop-walk round a stack, stopping every now and again to tilt its head and peer at the bottom, reminding me of a mousing cat. I don't know whether or not it caught anything on its foot-tour. At another derelict steading barn owls and jackdaws nested behind the lathing on the walls. In their last breeding season the owls took over a jackdaw's nest containing two eggs, but the six owlets hatched in it were killed by a feral cat. That summer somebody stripped the walls of their lathing and the owls and jackdaws departed.

The wood in which I did most of my owl-watching was, for many years, a haunt of the woodcock and year after year, from my tree seats, I could watch the roding flight of the males against the open sky above the canopy. All the birds I watched there flew the same circuit every night, and they flew it withershins, the English of which is anti-clockwise. The roding flight is fast, the wing-beats slow and owl-like. During the flight the bird croaks and chirps, sounding to my ears like *croak-croak-chissick*. Sometimes a bird will break its flight, dip down into the trees, and reappear with another. They will then chase each other round and round the tree trunks. After such a chase roding often stops for the night, but the roding male is just as

*opposite.* Skylark with young. Their habitat is moors, fields, marshes, sand-dunes and the nest, made of grasses, is in long grass on the ground.

likely to top the trees and resume his flight. When roding is over the birds move to their feeding grounds.

There should no longer be any argument about whether or not the woodcock sometimes carries its young. Monty and I flushed a bird from her nest and she dropped a chick at our feet when she was higher than our heads. That could have been because the chick was snagged in her feathers, as often happens with pheasants. The clincher for me came when Monty and I watched a woodcock leading three chicks across a main road early in the morning, leaving a fourth behind, crouched in the middle. In a few moments the bird flew back to it, and crouched over it as though to brood it. When she flew to the verge the chick was no longer on the road.

Another voice I heard every night when I was perched in the owl wood was the snipe's: *jick-jack: jick-jack: jick-jack.* On some nights, long after the owls were on the wing, I would hear a snipe drumming high overhead, a bleating sound produced by the tail feathers when the bird swoops steeply. Snipe nested in the rough, threshy field on the north side of the wood. So did peewits and curlews, and now and again one or other of them would fly into the woodcock's airspace when alarmed, probably by a fox or badger.

The nightjar, a swallow look-alike, although not like a swallow, sings and flies by night. For many years nightjars nested in a fern brake not far from one of the woods where I worked with owls, and on some evenings I came their way at dusk to see them before going on to spend most of the night up a tree. When I checked the hen on the nest the cock bird would be up and around, calling *kwik-kwik-kwik* until I withdrew to his toleration distance. The sitting bird matching her background so closely that she was difficult to see, and in certain lights the fact that I couldn't see her two eggs was the only way I could tell that she was there.

Nobody could possibly mistake the swallow-like nightjar for any other night-flying bird. Its churring song, too, is quite unmistakable. I heard it every night when I was treed with the owls and, on nights when the light was good, I could watch one or other of the pair hunting insects — flapping, gliding, wheeling and zigzagging in silent flight. I have heard the nightjar called fern owl and goatsucker. Swifts, also swallow look-alikes, become night flyers when they go high into the air to roost on the wing.

During the breeding season the Manx shearwater flies to its nest in the dark, and it is a great experience to visit a colony at this time. There are colonies from Orkney and Shetland to the Inner and Outer Hebrides, but I had never seen one on a mountain top until I visited Rhum with Jim Lockie after the then Nature Conservancy had taken over the island. The colony is on Hallival, among rocks and scree, a desolate place honeycombed by the burrows of shearwaters. We soon found a burrow in which a bird was sitting on an egg and lay down close by, hoping that this would be a night for the birds changing over. The shearwater will sit for two to four days at a stretch, sometimes even longer.

The day before we had watched the birds after sunset, flying low and fast above the waves, tilting from one side to the other, with the lower wing almost touching the water. They were waiting for darkness. Even in moonlight they are not keen to fly in from the sea. That night there was still a lot of light about. The afterglow had paled, but it hadn't faded out. It remained a slow-spreading stain behind the hills. A kind of twilight did; however, settle on our part of the mountain, and we hoped. And, suddenly, the birds were there, crooning and crowing, throaty and dove-like, over and round our heads, filling our ears with their cries, flickering like great bats, but swift and erratic as swallows. Birds alighting at burrows were greeted with crows and croons from birds in burrows. Hallival had become a tower of Babel.

As for the night flights and calls of the storm petrel and Leach's petrel, the late Sir Frank Fraser Darling, that bonny and caring man, said it all when he wrote: 'To be present at one of the breeding stations on a fine July night is one of the great experiences for a naturalist.' So it is. And for anyone else. Unfortunately, their breeding stations are mostly remote and not easily reached.

# TO CARE OR NOT TO CARE

The Grey Lag goose plucks down from its breast to line the nest. When it leaves the nest it covers the eggs with down, but crows get wise to this and poke amongst the down for eggs.

Wildlife is part of our national heritage. True. Wildlife is there for all of us to enjoy. Not true. Lots of people get enjoyment from wildlife, but that is not why it is there. It is there because it is there — after millions of years of evolution from other forms of wildlife which were there long before human beings appeared on the face of the earth. We don't know what exterminated the dinosaurs because we weren't around at the time. In his *Leaves of Grass* Walt Witman said it all when he described himself as an acme of things accomplished and an encloser of things to be.

I am one of those who believe that animals have rights; others take the view that they have none, that they are there for our convenience, to be exploited as we wish; and do. Most people draw the line at cruelty, but many don't, so animals have to be defended by activists from among those who draw the line.

It is a measure of our uncaring that gin-trapping and poisoning remained legal into the second half of the twentieth century, and strychnine can still be used to kill moles. Before gins were outlawed millions of wild animals died lingering, painful deaths in them, and many were left to rot in them. The same is true of poisons, and in 1980 I picked up two dead

*opposite*. A Woodpigeon feeding her half-grown young. Pigeons don't feed their young directly but produce a 'milk' in the crop and this is fed to the young.

173

rabbits laid out to poison eagles or buzzards, or anything else with a taste for rabbit. You'll still see cats moving around with the first joint of one forepaw missing, perhaps even both, and cats are not noted for chewing their feet off.

You will also hear it said that wildlife *needs* man: yes, in the same way as I need a bullet from ear to ear. It is we who have brought about the situation that makes the future of wildlife dependent on our changing our ways. Ours is the power but not very often the glory. Pope said:

> Know then thyself, presume not God to scan;
> The proper study of mankind is man.

Yes indeed; warts and all. Part of our self-study has to be an assessment of our trusteeship of the total environment; an examination of our role as the world's dominant species; an appraisal of our management or mismanagement of the world's resources: soil, water, forests, minerals, plants and wildlife. What is ecologically bad or unsound cannot be ethically or morally correct.

No one, and certainly not I, would deny our right to exploit and wisely use any natural resource. But to misuse or abuse, to squander or destroy, arbitrarily and arrogantly — these are a betrayal of trusteeship, an abdication of responsibility. Conservationists are often accused of being blindly opposed to progress whereas we are simply opposed to blind progress. Blind progress to date has resulted in man-made deserts and soil erosion, river and sea pollution and destruction of wilderness. We are constantly changing the face of nature in our own interest, which are in no way helped by destruction of habitats, despoliation and poisoning of the environment, or the elimination of wildlife.

What we lack is any overall strategy for the environment that will take into account the conservation of wildlife and habitats at the planning stage of any proposed development. Too often concerned voices are raised halfway through a project, or later, and by that stage crisis measures are required, usually too little too late. Twice I represented protesters, one of them a Local Authority, against proposed developments. Both hearings were held to consider whether or not planning permission should be given. In both cases planning permission was refused. On another occasion a different Local Authority asked me to check a pit bing for badgers before it went ahead with a proposed plan to demolish it. Two contractors, engaged in earth moving, roped off badger setts, at my suggestion, to prevent any damage by the earth-moving machines. They were all steps in the right direction.

Sound management of wildlife and habitats requires research, an understanding of how ecosystems work, the relationship of one species to another and their habitat, knowing when to protect, when to leave alone and if and when to cull, as in the case of the red deer and the roe. All too often conservationists and vested interests are on a collision course, when what is needed is consensus arrived at on the basis of fact. The go-getters and the tearaways and the fast-buck merchants all speak of conquering nature; the ecologist tries to understand and work in harmony with it.

What use is wildlife anyway? People sometimes ask the question, as they are entitled to do. A hundred years after Pope, the North American Indian Chief, Seathl by name, anticipated the question when he wrote to the President of the United States:

*What is man without the beasts? If all the beasts were gone, men would die from great loneliness of spirit; for whatever happens to the beasts also happens to man. All things are connected. Whatever happens to the earth befalls the sons of the earth.*

Seathl was an ecologist before the word was invented. And a savage, of course.

The Government's advisory body on the environment, the conservation and protection of wildlife and the countryside, is the Nature Conservancy Council. There are also a great many voluntary organisations engaged in the conservation and protection of wildlife, among them the Scottish Wildlife Trust, Friends of the Earth, and the Royal Society for the Protection of Birds. The fact that such bodies exist is a great reassurance to people generally; the sad truth is that they exist because they are necessary. Without them wildlife would have a much rougher time than at present. The Scottish SPCA was in the vanguard of such organisations and has demonstrated time and again its ability to adapt to changing circumstances.

But what a proud day it will be for all of us when we don't need them any more!

Fledgling thrushes perching
on the branches of a
flowering currant.

Jess,

Now I have to say *Hasta Luego*!, leaving you in the twilight that will shield you from distress at my going to face the temptresses on the frontier of darkest Nowhere. We have fought, and won, and been scarred by, many battles, you and I. This time we are fighting separate battles and, for the first time in our lives, can't help each other. Or can we?

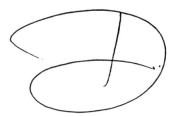

*Previous publications by David Stephen*

WATCHING WILDLIFE
SCOTTISH WILDLIFE
THE WORLD OUTSIDE
NATURE'S WAY (with James Lockie)

*Novels of Wildlife for adults and Children*

STRING LUG THE FOX
SIX POINTER BUCK
ALBA THE LAST WOLFE
BODACH THE BADGER
RORY THE ROEBUCK